ISO and ANSI Ergonomic Standards for Computer Products
A Guide to Implementation and Compliance

Wanda J. Smith

For book and bookstore information
http://www.prenhall.com

Prentice Hall PTR
Upper Saddle River, New Jersey 07458

Library of Congress Cataloging-in-Publication Data

Smith, W. J. (Wanda J.)
    ISO and ANSI ergonomic standards for computer products / by Wanda J. Smith.
       p.   cm.
    Includes index.
    ISBN 0-13-151119-X
    1. Computer software—Human factors—Standards. 2. Electronic digital computers—Standards. I. Title.
QA76.76.H85S65
004'01'9—dc20

1996
95-39031
CIP

Acquisitions editor: Bernard Goodwin
Cover designer: Wanda J. Smith
Cover director: Jerry Votta
Manufacturing buyer: Alexis R. Heydt
Compositor/Production services: Pine Tree Composition, Inc.

© 1996 by Prentice Hall PTR
Prentice-Hall, Inc.
A Simon & Schuster Company
Upper Saddle River, New Jersey 07458

All rights reserved. No part of this book may be reproduced, in any form or by any means, without permission in writing from the publisher.

The publisher offers discounts on this book when ordered in bulk quantities.

For more information contact:
  Corporate Sales Department
  Prentice Hall PTR
  One Lake Street
  Upper Saddle River, New Jersey 07458

  Phone: 800-382-3419
  Fax: 201-236-7141
  email: corpsales@prenhall.com

Printed in the United States of America
10 9 8 7 6 5 4 3 2 1

ISBN: 0-13-151119-X

Prentice Hall International (UK) Limited, *London*
Prentice Hall of Australia Pty. Limited, *Sydney*
Prentice Hall Canada, Inc., *Toronto*
Prentice Hall Hispanoamericana, S.A., *Mexico*
Prentice Hall of India Private Limited, *New Delhi*
Prentice Hall of Japan, Inc., *Tokyo*
Simon & Schuster Asia Pte. Ltd., *Singapore*
Editora Prentice Hall do Brasil, Ltda., *Rio de Janeiro*

This book is dedicated to my daughter Cindy, whose many birthdays I missed because I was at standards meetings.

Contents

Preface   xi

Chapter 1 Standards and Standards Organizations   1

Overview   1
Types of Standards   1
Ergonomics   3
Ergonomic Standards   5
Ergonomic Guidelines   8
Ergonomic Regulations   9
Standards Organizations   11
United Nations Organizations   20
European Standards Agencies   20
Ergonomics Associations   22
National Standards Agencies   24
United States Standards Agencies   24
Summary   34

Chapter 2 History of Ergonomic Computer Standards   35

Overview   35
Emergence of Ergonomic Standards   35
European Ergonomic Conferences   37
World Health Organization VDT Report   40
The United States Situation   41
The Ergonomic Standards Debate   42
Formation of ISO Ergonomic Standards   49
Europe's Solution   49
Ergonomic Legislation in the United States   50
Labor Union Activities   53
Summary   56

Chapter 3 ISO 9241   57

Overview   57
ISO 9241   57
Application Domains   75
Status   76
Content Expansion   76
Other Applications   77
Compliance   80
Summary   83

Chapter 4 European Ergonomic Requirements   84

Overview   84
The EU and EFTA   84
EEA   86
National Standards Agencies   86
European Standards   86
European Directives   89
EU Directives   89
CEN   94
Germany   95
Sweden   101
Denmark   106
Norway   106

Italy 107
United Kingdom 107
ESPRIT 109
Summary 110

Chapter 5 North American Ergonomic Standards 111

Overview 111
United States Standards 111
Canadian Ergonomic Standards 128
Summary 131

Chapter 6 U.S. Ergonomic Requirements for Special Circumstances 133

Overview 133
Requirements for the Disabled 133
Americans with Disabilities Act 135
Information Technology Access 138
Design Guidelines for GSA 508 141
Ergonomic Health Requirements 142
California Safety Standards 143
National Safety Ergonomic Standards 147
Summary 150

Chapter 7 Ergonomic Standards in Other Countries 151

Overview 151
Australia 151
Japan 157
Former Soviet Republics 161
Summary 162

Chapter 8 Ergonomic Checklists 163

Overview 163
Part 1: Basic Safety and Design Guidelines 164
Part 2: ISO 9241 Checklists 169
Part 3: Hardware Requirements 211
Part 4: Software Guidelines 217

Part 5: EU Directive Requirements   227
Part 6: Overview of Legal Requirements of the ADA   232
Part 7: Guidelines for User Interfaces for Individuals with Disabilities   233

Chapter 9   Usability Testing   238

Overview   238
Background   239
General Test Conditions and Procedures   243
Test Equipment   246
Monitoring Equipment   247
Test Procedures   248
Keyboard Test   250
Non-keyboard Input Device Tests   253
Biomechancial Assessment   258
Comfort Assessment   261
Display Usability Test   261
Software User Interface Tests   265
Summary   266

Chapter 10   The Impact and Future of Ergonomic Standards   269

Overview   269
The Ergonomic Explosion   269
Corporate Strategies   271
Product Strategies   272
Process Strategies   281
ISO 9000   284
The Future and Ergonomic Standards   287
Summary   288

Glossaries   290

Standards Agency Glossary   290
Standards Glossary   292
Standards Terms   295
Hardware User Interface Glossary   296
Software User Interface Glossary   306

Appendix	308
Bibliography	311
Index	325

Preface

At various times in my life people have told me that I seem to be having too much fun to be serious about my work and my profession. Participating in the development of standards is serious business, but it can also be fun. It requires a tremendous investment in time and other resources. In spite of this, I have always enjoyed it, met many very bright and dedicated professionals, and had some wonderful times while attending standards meetings. As with most international consensus processes, standards meetings are intense, often frustrating, and all-consuming. Perhaps that is why many of the associated social events are so enjoyable and memorable. They provide a welcome opportunity to unwind at the end of a long day of meetings.

One of the major personal benefits of my involvement in international standards activities has been the improvement in my technical and professional skills. Having to create design and measurement specifications that are continuously challenged during the standards development process has forced me to stretch my technical knowledge to new breadths and depths. Having to justify new and controversial standards requirements and metrics has helped improve my written and verbal communication skills. Working intensely to reach a consensus with people from different cultures, and with different values and agendas has increased my patience and understanding and improved my negotiation skills. I have seen similar changes in many other Americans who have become long-term members of international standards committees. It is amazing how consensus can finally be reached among people of different countries and beliefs when their efforts will not progress unless they can come to some sort of agreement. It is unfortunate that more Americans could not have the opportunity to participate in similar international events. I believe it would reduce American ethnocentricity and improve our image abroad. It could also contribute to world peace.

It is not very often that creators of standards receive thanks for the many evening and weekend hours spent in meetings. Most of us who are standards committee members have other jobs and many other professional and personal responsibilities. In addition, feedback to standards committees is often in the form of critique and complaint from those for whom standards may be burdensome. However, the hard work of resolving opposing views is at the core of the consensus-building process that results in better standards and, in the end, better products and work environments.

I would therefore like to take this opportunity to thank those who faithfully attended standards meetings, helped create the ISO and ANSI computer standards, and did so much to make the meetings and social occasions so enjoyable.

First, I would like to honor the memory of Etienne Grandjean, the father of ergonomics in Europe, who hosted me during many visits to the research center at the Technical Institute in Zurich. His research became the basis for many ergonomic standards and his contributions to computer ergonomics were revolutionary. I would like to thank Jim Greeson for his help with the ISO color display standard and all those who helped me with the input device standards for ISO and ANSI, especially Kathy Uyeda and Harold Welsh. I would also like to extend my appreciation to the various agencies who hosted our meetings, made endless copies of documents, and hosted luncheons and dinners. Special thanks go to BSI in London, AFNOR and IRRST in Paris,

DIN in Berlin, Olivetti in Everia, and SAS in Copenhagen. To Brent, Vincent Ahmet, Tom, and Tomas, I owe a special thanks because they made the long days in meetings bearable by lighting up the European night life.

I would like to extend special recognition to the following people for their contribution to the ISO VDT standard and its standard process:

- Peter Haubner, who was the convenor of the ISO display committee for over ten years and is well known for his excellent research on displays
- Gerd Dziambor, who was the next convenor of the ISO display committee and has conducted compliance testing since Germany created the first VDT standard
- Floris van Nes who conducts research on perception and whose diplomacy, charm and negotiating skills are an inspiration to us all
- Tom Stewart who without a doubt is the most knowledgeable person in the world on ergonomic standards and has probably done more to enhance ergonomic standards than anyone

I owe a special thanks to a number of people whose help and support made my trips away from home less lonely and stressful:

- Pete and Luda, who provided love and security through the years and cared for my animals and home during my standards sojourns
- Susi, Catherine, Olga, and Tove, who made the days between standards meetings special by accompanying me while skiing at St. Moritz, horseback riding at Chantilly, shopping in Paris, and sailing in the Swedish archipelago
- Brigette and Jean Charles, who made their Hotel Garvarni in Paris my "home away from home"
- Scott, Jill, and Tracy, my surrogate kids, who visited me in Paris, had their lives changed forever, and became the morning tabloid report at standards meetings

I also owe a great deal of gratitude to those who received the manuscript and gave me many valuable comments:

- Tom Stewart, a Scotsman, Chairman of ISO TC159/SC4/WG5 and Convenor of CEN TC 22 and ISO 9241
- Pat Billingsly, an American and ANSI/HFES 200 committee member
- Gerd Dziambor, a German and ISO 9241/SC4/WG2 Convenor
- Brent Duchon, an American and ISO/SC4/WG3 member
- Luda Toutolmin, a Russian who reviewed the manuscript by candelight while flooded-in during the storms of 1995

In invite the readers of this book to the world or ergonomics and its standards. I truly believe that if more people integrated ergonomics into their lives, the world would be a better place.

The intent of this book is to provide information on ergonomic computer standards, regulations for computer products, and their associated equipment and environments. It explains why these requirements exist, how they are developed, and their importance to the marketability of computer products. It also describes which computer standards and regulations are the most important to meet and how to meet them.

This book is primarily written for professionals for whom ergonomic computer standards have the most direct impact. These include:

- designers of computer hardware and software
- manufacturers and distributors
- employers
- unions
- users groups
- test agencies
- research institutes

Because ergonomic issues are being increasingly included in legislation, litigation, and quality assurance testing programs, the information presented in this book should also be of interest to:

- legislators
- attorneys and judges

- forensic consultants
- auditors and inspectors

Specifically, the contents of the book should help these professionals:

- understand the relevance of ergonomic standards and laws
- understand the scientific, political, legal, and economic aspects of ergonomic standards
- understand the international impact and focus of ergonomics and ergonomic standards
- know what types of ergonomic standards exist and their basic contents
- establish product design and use specifications
- evaluate product usability
- compare usability and compliance between products
- obtain information on ergonomic standards that is difficult to obtain from other sources

The book is composed of 10 chapters:

Chapter 1, **Standards and Standard's Organizations,** presents an overview of major ergonomic international and national requirements including standards, guidelines, and legislation. It describes the relationship between ergonomics and standards and the organizations that participate or influence ergonomic standard's development. It also provides an overview of the international and national U.S. standard development process.

Chapter 2, **History of Ergonomic Standards,** summarizes why and how ergonomic standards emerged nationally and internationally. Specifically, it provides a description of the concerns and issues that caused the emergence of ergonomic standards, the debate regarding their scientific basis, and the interest and involvement of national and international labor organizations.

Chapter 3, **ISO 9241,** provides an overview of the contents of the international ergonomic standard on computer displays, their workstations and environments. It summarizes the applications of this standard, describes how compliance is obtained and briefly summarizes its general contents.

Chapter 4, **European Ergonomic Requirements,** describes the EU and EFTA, their standards cooperation, European visual display terminal (VDT) standards, and the European ergonomic directive, its compliance and non-compliant penalties.

Chapter 5, **North American Ergonomic Standards,** describes computer ergonomic standards in the United States and Canada. It includes an overview of major U.S. military standards, a summary of the U.S. VDT hardware and software standards, and Canada's VDT standard.

Chapter 6, U.S. **Ergonomic Requirements for Special Circumstances,** describes U.S. regulations for the disabled: the *Americans for Disabilities Act* and *Access of Information Technology*. It provides a brief history and status report of several proposed computer design and use regulations for repetitive work at computer terminals.

Chapter 7, **Ergonomic Standards in Other Countries,** summarizes ergonomic computer standards in Australia, Japan, and the former Soviet Union. It describes some major differences between their emphasis on standard requirements and some background on how these differences evolved.

Chapter 8, **Ergonomic Checksheets,** provides several lists and tables summarizing product design requirements that are expected to dominate ergonomics for the next several years. It includes: a basic design checksheet, mandatory and recommended ISO requirements, software user interface guidelines and requirements, ADA legal requirements, and requirements for disabled individuals.

Chapter 9, **Usability Testing,** contains descriptions of user testing of computer hardware and software products for conformance with standards and regulations along with comparison testing and competitive analysis. It provides test procedures to assess user performance, comfort, and biomechanical load.

Chapter 10, **Impact and Future of Ergonomic Standards,** discusses how ergonomics has impacted and will continue to impact both product design and acceptance, and thus marketability. It includes examples of what is anticipated in ergonomics for the future of input devices, software, hardware, and work environments. It also suggests strategies that can help corporations maintain a competitive edge in ergonomics.

Glossaries are provided to help the reader sort out the terms used in the worlds of standards and ergonomics. The Bibliography summarizes the articles from which the information in this book was obtained.

This book was created on *Microsoft Word 6.0*. The icons were created using *Microsoft Powerpoint, Presentation Task Force, Grapic Works for Windows, Lifeart,* and *Corel Gallery*.

Although every attempt was made to provide accurate and up-to-date information, there are probably many topics and details that have been unintentionally omitted. The world of ergonomics and its standards, laws, directives, ordinances, and guidelines is confusing even to those who create and interpret them. In addition, ergonomic computer-product research results and requirements are evolving almost as fast as computer technologies. With such dynamic topics, it is difficult to publish up-to-date material in a timely manner. The material in this book thus represents information publically available at the time of its publication. The opinions expressed in this book are those of its author.

Chapter 1

Standards and Standards Organizations

Overview

This chapter presents an overview of ergonomic standards, regulations, and guidelines and how they are developed. It describes the role of some of the professional and industry organizations that create these standards. It also provides an introduction to international and national standards and specific information on the ergonomic standards of the International Organization for Standardization (ISO) and the American National Standards Institute (ANSI).

 **Types of Standards**

A standard is a requirement, rule, or recommendation based on proven principles and practice. It is established by a recognized authority for the measurement of quantity, weight, extent, value, or quality. A standard represents an agreement by a group of officially sanctioned professionals at either a local, national, or international level.

A *local* or regional standard is an accepted design or practice of a trade, professional organization, or business entity; a *national* standard is a convention acccepted by a wide variety of organizations within a nation; an *international* standard represents a consensus among standard organizations around the world.

The primary objectives of standards are to:

- provide safe and healthy products and environments
- promote economy of human effort
- establish product/process consistency
- protect consumer interests
- promote quality of life
- protect the environment
- promote trade by removal of barriers caused by differences in national practice

In addition to their general benefits to society, standards provide several business advantages. They benefit both employers and employees by motivating the integration of safety in product design. They improve production cost, effort, and safety by minimizing errors and delays in operation. They increase usefulness of products and thus extend their applications and marketing possibilities.

Standards can be either *normative* or *non-normative*. *Normative* standards contain both mandatory and optional requirements; *non-normative* standards are strictly optional or voluntary. Normative mandatory requirements are indicated by the use of "shall" in their specifications; normative optional requirements are indicated by the use of "should." Non-normative standards use only "should."

Standards can also be either mandatory or voluntary. Mandatory standards must be met; voluntary standards are optional. Thus, product manufacturers are not required to comply with a voluntary standard, but if they claim compliance, they must comply with all requirements in the standard. For example, the American National Standards Institute's ergonomic standard for computer workstations (ANSI/HFS 100, see Chapter 5) is a voluntary standard. Manufacturers who do not meet all of its requirements cannot claim full compliance.

In addition to standards, there are several other types of product and environmental requirements, including laws, directives, guidelines, and technology agreements. Standards can be created by international, national, and local government or private agencies; laws are created by government bodies; directives are created by multinational governing organizations; guidelines can be created by agencies or individuals; technology agreements are created between companies. Although standards are not laws, they can become legislated. Standards can also be used as evidence in litigation proceedings (see Chapter 10: Impact and Future of Ergonomic Standards).

There are a variety of entities for which standards are developed, including products, equipment, technologies, environments, processes, and measurement. The general public is probably most familiar with consumer product safety standards. Less well known, but having an increasingly important role in product design and use are ergonomic standards. In order to provide an understanding of ergonomic standards, ergonomics needs to be described.

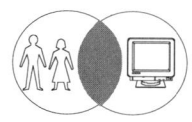 **Ergonomics**

Ergonomics is the study of human capabilities and limitations, human interaction with technologies and environments, and the application of this knowledge to products, processes, and environments. The field of ergonomics is based on human perception, cognition, anatomy, biomechanics, physiology, and linguistics. The following are examples of ergonomic research applied to a variety of products and situations:

- evaluating the ability of users to identify and respond to tool bar icons in a computer display (perception research)

- evaluating the ability of users to determine and remember the meaning and results of computer commands (cognition research)
- evaluating the ability of users of different sizes and physical limitations to fit and efficiently accomplish tasks within a computer workstation space (anthropometric research)
- determining muscle load resulting from holding and activating input devices (biomechanics research)
- measuring indicators of visual functioning (like blink rate, acuity, and accommodation) to determine the stress effects of alternately viewing white paper and a black background on a computer screen (physiology research)
- evaluating the appropriateness of syntax and semantics of a software user interface (linguistics research)

The application of ergonomics also requires knowledge of a diversity of fields, such as product and environmental design and function, work environmental effects, experimentation, statistics, physics, and advanced mathematics. A knowledge of product and environmental design is necessary because the majority of ergonomic engineers evaluate the appropriateness and quality of product (and environments) in terms of work effectiveness, usability, health, safety, and comfort. These evaluations require not only a knowledge of products and the ways in which they may be used but also the characteristics of their users. In addition, at some time in their careers, most ergonomists are requested to conduct a study or controlled test of product usability. This requires the skills of designing a valid and reliable test, which include how to:

- select an appropriate subject population sample
- determine appropriate test tasks
- use performance monitoring equipment
- design data collection metrics
- utilize appropriate statistics to analyze test data

A knowledge of physics is used for the analysis of stress such as that imposed on users by the mechanical forces of lifting or keyboarding. A knowledge of integral calculus is used in determining muscle fatigue and perception of flicker.

The following are just a few aspects of computer products that are determined by the application of a wide range of knowledge:

- labeling and layout of keys
- symbols to represent menu commands
- shape and force of mouse and puck buttons
- weight of lap top computers
- type and number of commands in a menu hierarchy

In the United States, *ergonomics* was historically known as *human factors engineering,* and the focus of its professional activity was on assessing work efficiency. In Europe, the term *ergonomics* was traditionally used, and the focus of European ergonomic activities was on assessing human efficiency, safety, and health. Due to the socio-political influence of Europe on Japan, India, Australia, and New Zealand, the activities of these countries have also focused on ergonomics rather than human factors engineering. The global influence of Europe in ergonomics and ergonomic standards has resulted in the widespread use of the term *ergonomics* throughout the United States. In addition, the human factors professional organization, which is headquartered in the United States, recently changed its name from the Human Factors Society (HFS) to the Human Factors and Ergonomics Society (HFES).

 ## Ergonomic Standards

Creators of Ergonomic Standards

Ergonomic standards are typically created by a committee of professionals trained in human factors engineering/ergonomics and such related fields as industrial hygiene, mechanical engineering, industrial design, and computer science. Committee members are appointed by a standards agency or an organization authorized by a standards agency.

Not all ergonomic standards are created by national and international agencies. Some are the result of tradition, advertising, and market domination and are referred to as *de facto standards*. Examples of de facto standards for software user interfaces are the Microsoft Windows™, Apple Macintosh™, and Motif™ operating environments and hardware user interfaces like the QWERTY keyboard layout and the calculator keypad layout.

Since standardization results in consistency in design and testing, it is important that product design be ergonomically optimized before it is standardized. Ergonomic design optimization can be achieved if ergonomic principles and practices are incorporated both in product design and in testing processes (see Chapter 10).

Types of Ergonomic Standards

Ergonomic standards include the specification of design, testing, and use of products as well as environmental conditions. Ergonomic standards are based on principles derived from ergonomic analysis. There is a fundamental difference between a technical standard and an ergonomic standard. A *technical standard* specifies materials and artifacts as they are designed and produced by humans. An *ergonomic* standard specifies the characteristics of products as they are used by humans. Although most ergonomic standards are voluntary, they can be virtually mandatory in terms of a product's marketability. In the United States, non-compliance with voluntary ergonomic standards can be used as evidence in litigation proceedings.

There are a wide variety of ergonomic standards. For example, there are ergonomic *safety standards, product design standards,* and *work environment standards*. There are also ergonomic *standards for special technologies* and *special conditions* or *situations,* such as the proposed standards for flat panel displays (see Chapter 3) and repetitive tasks (see Chapter 6).

Ergonomic standards are national, international (see Table 1–1) and regional (see Chapter 4). Examples of national stan-

Table 1–1 Examples of national and international ergonomic standards

National
U.S. NIOSH: *Work practices for manual lifting*
U. K. BSI 381: *Colors for identification, coding, and special purposes*
Sweden SS01: *Color notation system*

International
ISO 3409: *Passenger cars—Lateral spacing of foot controls*
IEC 73: *Colours of indicator lights and push-buttons*
CIE 15.2: *Colorimetry*

dards for visual display (see Table 1–2) include the American National Standard Institute ANSI/HFS 100, the German (Deutsch) Institute of Normalization DIN 66234, and the Japanese Institute of Standardization JIS 6041. An example of an international standard is ISO 9995 (Keyboards); an example of a multinational standard for VDT workstations is CEN EN29241.

In addition, there are civilian and military ergonomic standards, which can often influence each other. For example, the U.S. Military Standard (MIL STD) 1472 is used for the procurement of military and government equipment (see Chapter 5). It emphasizes performance, reliability, and safety issues and requires that ergonomics be considered in the design of military equipment and systems. However, many of its requirements are used in civilian standards and are often proposed for inclusion in international standards as well.

Just as military standards can become the basis for civilian standards, one nation's standards can also be used as a basis for standards in other nations as well as for international standards. For example, many of the requirements of the U.S. MIL STD 1472 were integrated into the ANSI/HFS 100 standard for computer workstations and have been used in standards of other na-

Table 1–2 Examples of national and international VDT standards

National

U.S. ANSI/HFS 100:
Human factors engineering of visual display workstations
Canada. CSA Z412:
Office ergonomics
Germany. DIN 66234:
VDU work stations

International

ISO 9241:
Ergonomic requirements for office work with visual display terminals
CIE 60:
Vision and the visual display unit
ISO 9995:
Keyboard

tions. Likewise, many of the specifications in the German DIN 66234 visual display workstation standard have been integrated into other national standards. The international ergonomic standard for office work with VDTs (ISO 9241) is being regionally adopted by the European Union (see Chapter 4).

Ergonomic Guidelines

Ergonomic standards may also be in the form of guidelines. Guidelines are based on a variety of factors including:

- empirical data
- cultural expectations
- models of human behavior
- expert opinion

Most software user interface standards and specifications are currently in the form of guidelines (see Chapters 3 and 8).

Unlike mandatory normative standards, which are usually applied without interpretation or modification, guidelines (and non-normative standards) require judgment in their application. Guidelines are often included in non-normative standards and are usually indicated by use of terms such as "should," "may" or "can."

Ergonomic guidelines are typically created by business organizations, organized labor, and user groups before standards are available or when existing standards are insufficient. Both ergonomic standards and guidelines are often used in labor negotiations and arbitrations. Guidelines are voluntary, but like standards may become requirements in certain markets or specific regions (such as the Swedish guideline *Screen Checker,* see Chapter 4).

The *Screen Checker* demonstrates how a guideline can have as much international influence as a standard. In the mid 1980s, a Swedish labor union created this guideline (see Figure 1–1), which specified design requirements for screens and keyboards, as well as heat, noise, and radiation emissions (see Chapter 4). Many of the specifications in this guideline soon became the basis for standards and labor negotiations in other countries. Another influential guideline is the *Guidelines for Designing User Interface Software* produced by Mitre Corporation (see Chapter 5). These guidelines have been integrated into national, international, and corporate guidelines and standards.

 **Ergonomic Regulations**

Regulations are legally mandatory, and thus are the most strict of all requirements. Non-compliance with a regulation can result in a range of ramifications and penalties from loss of business to monetary fines and a jail sentence (see Chapters 4 and 6).

Directives are legally binding instruments for implementing policies or decisions in the European Union (formerly the European Community, see Chapter 4). Directives are instructions from the European Council to one or more member states requiring them to legislate on a specific matter within a defined period of time. Directives provide outlines of required legislation but leave the details of implementation to the European member states.

As with other types of ergonomic requirements, regulations and directives can apply to both product manufacturers and employers. However, U.S. regulations for the design and use of office products (see Chapter 5) and the European Union (EU) Directive *Minimum Safety and Health Requirements for Work with Display Screen* (European Community, 1990) apply only to employers' equipment (see Chapter 4).

As opposed to standards, ergonomic regulations are usually not as technically detailed (see Chapters 3 and 4), and interpretation of requirements is left to local agencies.

Most countries have ergonomic standards for VDTs; some countries have legislation (see Chapters 4, 6, and 7); other countries have national labor union guidelines (see Table 1–3); some have all three.

Table 1–3 Countries with computer terminal requirements

Area & Country	VDT Ergonomic Requirements		
	Standard	Legislation	Labor Guidelines
Europe (EU)	✓	✓	
Germany	✓	✓	✓
Sweden			✓
France	✓	✓	
United Kingdom	✓	✓	✓
Netherlands	✓	✓	
Austria	✓		
Norway	✓		
Finland	✓		
Belgium	✓		
Denmark		✓	
North America			
United States	✓		
Canada	✓		
Pacific Rim			
Japan	✓		
Australia	✓		✓

Standards Organizations

Many groups play a role in creating or influencing the creation of ergonomic standards. These include international, national, and local institutes, organizations and agencies from industry, labor, and government (see Table 1–4). There are also a number of multinational standards organizations such as the European Computer Manufacturers Association (ECMA) and the Center for European Norms (CEN) that create standards.

International Organization for Standardization

The most influential and largest voluntary group of industrial and technical cooperation in the world is the International Organization for Standardization (ISO). ISO is a worldwide federation that develops, coordinates, and promotes international standards with national standards organizations from 90 countries. ISO was created to promote standards that could facilitate international exchange of goods and services and cooperation in the areas of intellectual, scientific, technological, and economic activity. ISO coordinates the interests of product manufacturers and users in the preparation of international standards. Although ISO is a nongovernmental entity, more than 70 percent of its membership are

Table 1–4 Examples of organizations involved in ergonomic standards

Organizations	National	International
Standards agencies	ANSI, ASTM, DIN, BSI, CSA, JIS, AFNOR	ISO, IEC, CENELEC, CCITT
Labor organizations	AFL CIO, TCO, TUC	ILO
Industry agencies	AEA, ITI, VDMA	
Health agencies	OSHA, NIOSH	WHO
Professional organizations	ACM	IEA, CIE, HFES, IEEE
Government agencies	DOD, DOT, DOE	

governmental agencies or bodies that are incorporated by public law.

The objective of ISO is to insure that standards are scientifically reliable and meet specific requirements. The scope of an ISO standard is not limited to any particular field or profession but includes all fields except electrical and electrical engineering. Standards for these fields are the responsibility of the International Electrical Commission (IEC) and the Commission of Illuminating Engineering (CIE).

ISO technical work is conducted by more than 180 technical committees (TCs) and 1,500 working groups (WGs). The result of ISO technical work is ultimately published in the form of International Standards (ISs) and Technical Papers. There are more than 7,100 ISO standards; they are usually updated every five years or when deemed appropriate due to technical developments.

ISO Standard-Development Process. The development of an ISO standard occurs in several stages (see Table 1–5). The first stage is the creation of a *work item* (WI). In order for a potential standard topic to become a work item, it must be supported by at least five countries and submitted to ISO for standard development approval. Once a work item is approved, ISO requests representatives from national standards organizations of ISO member countries to form a work group (WG) to assemble information for the new standard.

After the work group specifies and consolidates requirements, the committee votes to determine if the work item is to become a *committee draft* (CD) proposal. The CD stage is the first formal voting stage of a standard. It is reviewed by participant (P) members with voting rights and observer (O) members who receive standard drafts for information only. If the CD is not approved by a majority of voting members, it is returned to the working group for review of the members' comments and revision. When the CD is sufficiently revised to recommend advancing it to a *draft international standard* (DIS) status, the document is sent to interested ISO members for a vote.

If the CD is approved, it advances to a draft international standard (DIS) status. The DIS is revised by the working group, which reviews and incorporates changes resulting from country member comments. Major technical changes are not supposed to

Standards Organizations

Table 1–5 Stages of development of an ISO standard

Stage	Stage Title	Description
1	*Work Item (WI)*	An approved and recognized topic addressed by a working group (WG) that can result in one or more published standards. Documents are circulated within the specific work group. Committee agreement is required to advance a WI to a CD.
2	*Committee Draft (CD)*	A document circulated for comment and approval within the working group and national committees that mirror the work of other ISO committees. Voting approval is required by a majority of national members for the draft to reach the next stage.
3	*Draft International Standard (DIS)*	A draft standard is widely circulated for public comment via national standard bodies. Voting approval of a majority of member countries is required for the draft to reach the next and final stage.
4	*International Standard (IS)*	The final published standard. Documents are available from national standard bodies or from ISO headquarters in Geneva. After a DIS is approved as an IS, Geneva will publish the standard in English and French. The publishing process takes about one year.

occur at the DIS stage unless requested by a majority of voting members. When revision of the DIS is completed, it is again sent to national standard organizations (such as ANSI, DIN and JIS) for final vote to IS status. Each national standard agency that is a P member can contribute one vote.

The DIS is the second formal voting stage for an international standard. If the DIS is approved by a majority of member countries, the document becomes an international standard (IS); if not, the DIS is returned to the working group for review and revision. If it is not approved as an IS after a second vote, the DIS reverts back to a CD.

ISO Ergonomic Standards Committee. The ISO technical committee (TC) responsible for ergonomics is ISO TC 159. In general, its scope is to promote the adaptation of working and living conditions to the anatomical, physiological, and psychological characteristics of humans in relation to physical, sociocultural, and technological environments. Specifically, the standards development activities of ISO TC 159 are to:

- promote safety, health, effectiveness, and well-being
- request assistance from all ISO-member bodies producing standards that pertain to the work of ISO TC 159
- obtain specialists and experts for working groups
- ensure a balance of scientific competence and experience among experts to ensure that the contents of the standard are valid
- adjust the nature and scope of the contemplated tasks to the working group's size and competencies
- be aware of the consequences of existing laws, regulations, and codes of practice on proposed ergonomic standards
- foresee objections likely to be raised by social partners
- encourage and support members in raising funds for attendance at meetings and for the validation of the developed standards

The objectives of ISO TC 159 are to:

- collect and analyze ergonomic data pertinent to
  - the design and manufacturing of machinery
  - the design and organization of work processes
  - the layout of equipment and the workplace environment
- identify branches of industry, services, and trade where ergonomic needs will expand or arise with new technologies
- incorporate into standards development time lags in producing and enforcing international standards
- initiate and implement comprehensive standardization activities in the different fields of ergonomics
- make effective use of liaisons with other technical committees and subcommittees which may be using or are creating standards with insufficient consideration of available ergonomic data and/or principles
- create a strategy planning function in charge of implementing strategic policies

Table 1–6 ISO TC 159 subcommittees

SC	Subcommittee title	Scope
1	*Ergonomic guiding principles*	Standardization of basic principles and guidelines in ergonomics
2	*Ergonomic requirements to be met in standards*	Promotion and application of ergonomics in standardization activities of other ISO/TCs through consultations with, and recommendations to, TCs and SCs; preparation of ergonomic supplements to standards
3	*Anthropometry and biomechanics*	Provision of basic standards and guidelines for standardization in anthropometry and biomechanics with respect to terminology, methodology, data, etc.
4	*Ergonomics of human system interaction*	Standardization according to ergonomic principles in the field of acoustical and visual signals, visual displays, and their respective arrangements
5	*Ergonomics of the physical environment*	Standardization of terms, methods, and data applications for ergonomics design and evaluation of the physical environment of humans in their various activities
6	*Postures and dimensions for the design of workplaces at control stations*	Standardization of aspects of control stations that allow accommodation to postures and anthropometric considerations of users

Composition of ISO TC 159. ISO TC 159 is composed of six subcommittees (SCs) (see Table 1–6).

The ISO TC 159 subcommittees are divided into several working groups (WGs) (see Figure 1–1). The objectives of each working group is to enable users to perform their tasks under ergonomically favorable conditions.

The working groups that have created the most influential ISO ergonomic standards for computer products, workstations and their environments are WG2, WG3, and WG5.

WG 2 has created and continues to be in the process of creating ISO standards for:

- displays and display images (ISO 9241, Part 3)
- display color (ISO 9241, Part 8)

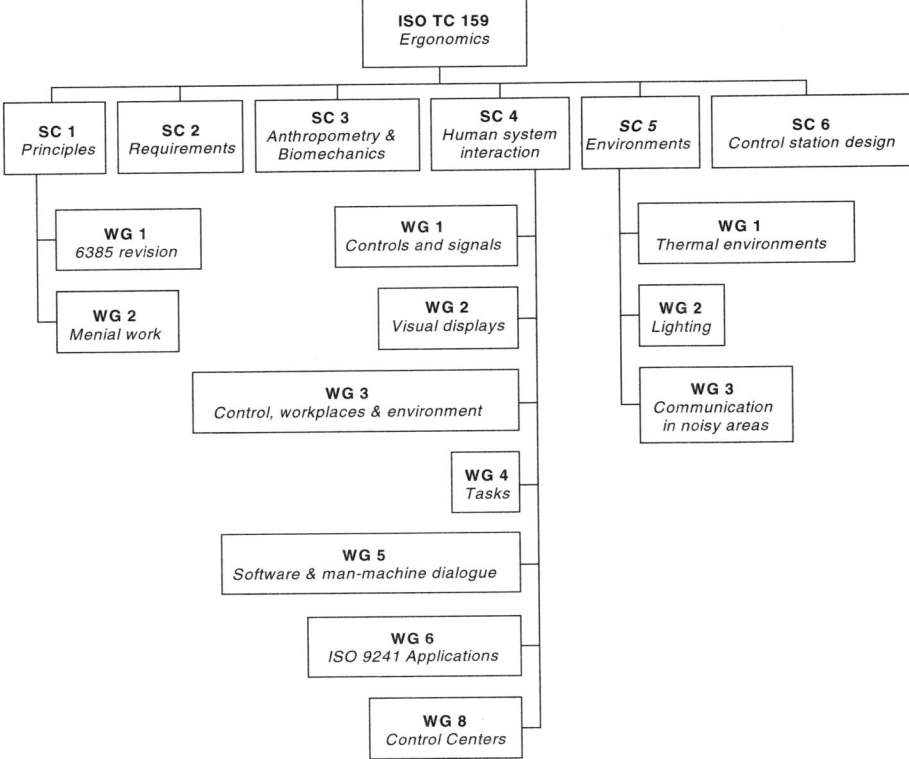

Figure 1-1 ISO TC 159 organization by topic

- screen reflections and glare (ISO 9241, Part 3)
- flat panel displays (ISO 13406-2)

WG 3 is creating ISO standards for:

- workplaces (ISO 9241, Part 5)
- environment (ISO 9241, Part 6)
- keyboards (ISO 9241, Part 4)
- non-keyboard input devices (ISO 9241, Part 9)

WG 4 developed Part 2 of ISO 9241 which addresses the task and job design aspects of work with VDTs.

WG 5 has created eight parts of ISO 9241:

- Dialogue principles (ISO 9241, Part 10)
- Guidance on specifying and measuring usability (ISO 9241, Part 11)
- Presentation of information (ISO 9241, Part 12)
- User guidance (ISO 9241, Part 13)
- Menu dialogues (ISO 9241, Part 14)
- Command dialogues (ISO 9241, Part 15)
- Direct manipulation dialogues (ISO 9241, Part 16)
- Form-filling dialogues (ISO 9241, Part 17)

WG 6 is working on guidance of the application of ISO 9241 to areas other than the office, the application of ISO 9241 to the design process, and revision of ISO 9241, Part 1 (see Chapter 3).

 International Electrical Commission

The organization responsible for preparing and publishing international standards for the electrical and electronics field is the International Electrical Commission (IEC). Founded in 1906, the IEC is a nongovernmental organization composed of National Committees in 44 countries. The work of the IEC is conducted by 88 technical committees (TCs), more than 100 subcommittees (SCs), and several hundred working groups (WGs). Each group is responsible for developing standards for a specific sector of technology.

The IEC cooperates closely with ISO and maintains working relationships with over 200 international governmental and nongovernmental bodies that are interested in international standardization in its fields. As one of the official reviewers of ISO ergonomic standards, the IEC reviews standards within its scope of interest.

  Harmonization

Harmonization of ISO and IEC standards is the responsibility of a joint technical committee (JTC). The joint technical committee for standards in the field of information technology is JTC 1. The subcommittee of JTC 1 working on standards for *Text and Office Systems Applications* is SC 18. The working group developing standards for user system interfaces and symbols is WG 9. Its work is divided into three major areas: 1) keyboards, 2) dialogue interaction, and 3) symbols and icons.

1. The keyboard standard (ISO/IEC 9995, see Chapter 3) development includes consolidation, harmonization, and update of existing keyboard standards and development of new keyboard standards taking into consideration new uses, technology, and users.
2. The dialogue interaction standard development addresses cursor control, scrolling, menus, forms, windowing, direct manipulation, and commands.
3. Symbol and icon standard development includes defining icon semantics and specifying icon shapes for text and office systems for software interfaces. Six parts are being created to include different kinds of icons: interface, object, printer, control, task, and status.

 Commission of Illumination Engineering

The Commission of Illumination Engineering (CIE) is an organization devoted to international cooperation and exchange of information among its member countries in all matters relating to the science and art of lighting. There are 32 member countries in the CIE and voting is limited to one vote from each country. In the United States, the CIE is represented by the U.S. National Committee (USNC-CIE).

The objectives of the CIE are:

♦ provide an international forum for the discussion and exchange of information relating to lighting science, technology, and art

Standards Organizations

- develop basic standards and procedures of metrology
- provide guidance on the application of principles and procedures in the development of international standards and national standards
- maintain liaison and technical interaction with other international organizations

The CIE operates through technical committees that address:

- photometry and radiometry
- colorimetry
- color vision
- visual signaling
- materials
- color rendering

The CIE plays an important role in providing technical advice to ergonomic standards such as the ISO standards for displays (such as ISO 9241—Parts 3, 7, and 8—see Chapter 3) and ISO 13406. CIE luminance and color definitions and measurement metrics are incorporated into all these standards.

 International Telegraph and Telephone Consultative Committee

The International Telegraph and Telephone Consultative Committee (CCITT) creates standards and recommendations for its members, which include the postal, telegraph, and telephone authorities (PTTs) around the world. The subgroup responsible for developing the standards for languages and methods for telecommunication applications is SGX. The working group responsible for man-machine interface for management of telecommunications is WP1. The standards of the CCITT are becoming increasingly relevant to computer workstations linked to telecommunication devices.

 ## United Nations Organizations

There are several global organizations that have significant impact on the creation and contents of ergonomic standards and regulations. The main ones are the World Health Organization (WHO) and the International Labor Organization (ILO)—both are special agencies of the United Nations.

Among other activities, the WHO promotes health standards and creates health and research guidelines. The primary objectives of the ILO are to improve labor conditions and living standards and to promote social justice and protect foreign workers.

Although the WHO and the ILO do not create standards, they often initiate and influence their creation and contents, especially by creating guidelines and publishing technical reports. For example, in 1985 WHO published a report on *Visual Display Terminals and Workers Health* (WHO, 1985). This document influenced the content of the EU *Display Screen Directive,* which became effective in January of 1993 (see Chapter 4). In 1994, the ILO published a guideline on working with visual display terminals.

In addition, the WHO and the ILO significantly impact public opinion toward product acceptance and use. They can thus influence global marketing demands and product success.

 ## European Standards Agencies

CEN and CENELEC

Some standards agencies are multinational but regionally limited. For example, Europe's two major standard organizations are the Center of European Normalization (CEN) and European Committee for Electrotechnical Standardization (CENELEC) (see Chapter 4). They are private sector organizations whose primary purpose is to develop standards that will enhance the harmoniza-

European Standards Agencies 21

tion of European interests and prevent technical and economic trade barriers. CEN TC 122 is responsible for harmonizing European ergonomic standards. Both CEN and CENELEC are committed to adopting standards developed by ISO and the IEC. For example, CEN TC 122 WG 5 is planning to adopt ISO 9241 as a CEN standard (see Chapter 4).

Multinational European agencies that affect standards and create guidelines include *economic* organizations such as the European Union (EU) and the European Free Trade Association (EFTA), *professional* organizations such as the European Telecommunications Standards Institutes (ETSI) and *trade associations* such as the European Computer Manufacturers Association (ECMA). Of the four organizations, ETSI is the only standards making body.

ETSI

The European Telecommunications Standards Institute (ETSI) creates standards for telecommunications technologies. Its members are national telecommunications administrators, network operators, manufacturers, and users in Europe. It has twelve technical committees, one of which deals with human factors (TCHF). There are four work programs regarding ergonomics in TCHF:

1. Telecommunications Services (STC HF1)
2. Usability Evaluation (STC HF3)
3. People with Special Needs (STC HF4)
4. Human Factors Support (STC HF5)

The work of Telecommunications Services includes ergonomic topics such as hardware and software controls and indicators, terminal user interface aspects, and phone based interfaces. The Usability Evaluation work program addresses usability measurement and testing for telecommunication terminals. The work group concerned with the People with Special Needs program is creating standards for telecommunication products. These include standards for public telephone access, keyboards, displays, and system response. The Human Factors Support work group is creating a *Human Factors Telecommunications Report and Handbook* and a glossary of definitions and terms regarding ergonomics in telecommunications.

ECMA

The European Computer Manufacturers Association (ECMA) is an industry consortium composed of 40 companies that manufacture computer products in Europe. ECMA publishes technical reports and standards that its member companies pledge to support and that are intended to influence European and ISO standards.

The ECMA technical committee on ergonomics is TC-28. Since 1980, it has published seven ergonomic standards and guidelines:

1. *Ergonomic Recommendations for VDU Work Places* (1982)
2. *Ergonomics Checklist for VDTs and VDT Workplaces* (1983)
3. *Ergonomics—Requirements for Monochromatic Visual Display Devices* (1985)
4. *Ergonomics—Requirements for Color Visual Display Devices* (1986)
5. *Ergonomics—Requirements for Non-CRT Visual Display Units* (1989)
6. *Procedure for Measurement of Emissions of Electric and Magnetic Fields from VDUs from 5 Hz to 400 Hz* (1991)
7. *User Interface Taxonomy* (1992)

These standards and guidelines have influenced several national and international ergonomic standards.

 **Ergonomic Associations**

The International Ergonomics Association

The International Ergonomics Association (IEA) is composed of ergonomics and human factors societies around the world. The IEA promotes the knowledge and practice of ergonomics by initiating and supporting international activities and cooperation. Its goals include the advancement of knowledge, information ex-

Ergonomic Associations

change, and technology transfer related to ergonomics. It attempts to accomplish its objectives by:

- providing and facilitating opportunities for international contacts among ergonomists
- cooperating with international organizations to facilitate the practical applications of ergonomics in industry and other areas
- encouraging scientific research by qualified persons in the field of ergonomic study
- sponsoring international ergonomic conferences

The IEA has 29 affiliate societies representing about 16,000 ergonomists worldwide (see Table 1–7). It maintains liaisons with the WHO, ILO, and ISO.

The Human Factors and Ergonomics Society

The Human Factors and Ergonomics Society (HFES) is an interdisciplinary professional association of individuals involved in the field of human factors engineering and ergonomics. It is part of the IEA and its membership includes over 5,000 professionals

Table 1–7 Countries with IEA federated societies

Europe		Americas	Pacific Rim	Near/Far East
Austria	Italy	Brazil	Australia	China
Belgium	Netherlands	Canada	Japan	India
Croatia	Poland	United States	Korea	Israel
Czech Republic	Portugal		New Zealand	Russia
Hungary	Serbia			Countries of South East Asia
France & Belgium	Slovakia			
Germany	Spain			
Greece	United Kingdom			
Denmark, Finland, Norway, & Sweden				

from all over the world. The HFES promotes the discovery and exchange of human factors/ergonomic knowledge. It is sanctioned by ANSI to create ergonomic standards for computer hardware and software user interface, related furniture, and the environments in which they are used (see Chapter 5).

The HFES publishes the technical journal Human Factors, hosts an annual conference on human factors engineering/ergonomics, and sponsors several other national and international conferences.

National Standards Agencies

National standards agencies officially authorize standards-making bodies in individual countries. They develop standards to meet their specific economic and social needs. Examples of national organizations that are active in creating ergonomic standards are ANSI (United States), DIN (Germany), and JISC (Japan) (see Appendix). Examples of national standards organizations that have created VDT ergonomic standards or ordinances are shown in Table 1–8.

United States Standards Agencies

In the United States, national standards bodies that create ergonomic standards are government agencies or organizations composed of professional members (see Table 1–9). Government agencies create ergonomic standards for the general public (such as the Occupational Safety and Health Administration), for the military (such as the Department of Defense), for public transportation (such as the Department of Transportation), and for federal staff (such as the General Services Administration). Professional agencies that create or oversee the creation of ergonomic standards for general public users of products and environments include the American National Standards Institute (ANSI), the Institute of Electrical and Electronic Engineers (IEEE), Association for Computer Machinery (ACM), and National Institute of Standards and Technology (NIST).

United States Standards Agencies

Table 1–8 Examples of standards agencies and VDT ergonomic standards

Country	Agency	Standard
EUROPE	CEN	EN 29241: *Ergonomic requirements for office work with visual display terminals*
Austria	ONORM	2630
France	AFNOR	X35-001: *Design of work systems, Ergonomic principles to be met*
Germany	DIN	66234: *VDT workstations*
The Netherlands	NEN	3002: *Ergonomic requirements for the design of visual display units and their input devices*
Norway	NSF	*Rules and recommendations concerning work stations at terminals*
Sweden	SSD	Directive 136: *Reading of display screens* AFS: *Ordinance concerning work with computer displays*
United Kingdom	BSI	7179
PACIFIC RIM		
Japan	JISC	6041: *CRT Display and Keyboard Units for Business Use*
Australia	SAA	*Screen Based Workstations*
AMERICAS		
United States	ANSI	HFS/100: *American National Standard for Human Factors Engineering of Visual Display Terminal Workstations*
Canada	CSA	Z412-M89: *Office Ergonomics*

Table 1–9 Examples of U.S. ergonomic standards bodies

Government	Professional
OSHA	ANSI
DOD	IEEE
DOT	ACM
GSA	NIST

U.S.–ANSI

The American National Standards Institute (ANSI) is the private membership organization that coordinates U.S. voluntary consensus standards activities. Its members include representatives of 1,100 companies, 250 technical, trade, labor, and consumer organizations, and 30 governmental agencies. ANSI is the official U.S. representative to ISO and IEC. The following information is obtained from ANSI brochures.[1,2]

ANSI Activities. ANSI does not directly develop or interpret American standards. It ensures that a single set of non-conflicting American national standards are developed by ANSI-accredited standards-developers. ANSI official activities include the following:

- *coordination of voluntary standards activities* ANSI assists standard developers and users from the private sector and government to reach agreement on the need for standards and to establish priorities in their development. ANSI also accredits qualified organizations to develop standards, helps them avoid duplication of effort, and offers a neutral forum for resolving differences. These functions are governed by the Executive Standards Council and standards boards.
- *approval of U.S. national standards* ANSI administers the only recognized system in the U.S. for establishing national standards. Its approval uses a consensus process, and its procedures ensure that any interested party can participate in a standards development or comment on its provisions. These due-process practices guarantee that American national standards maintain credibility and achieve broad acceptance.
- *representation of U.S. interests in international standardization* International standardization has become increasingly important in view of the changing global economy and European Commission initiatives in Europe (see Chapter 4). ANSI, on behalf of the U.S. voluntary standards system, influences the development and content of international standards to meet current technological needs and to facilitate competitiveness in the marketplace.

   ANSI helps govern ISO and IEC through its membership on their policy-making councils. ANSI participates in

United States Standards Agencies

most of the technical programs of both organizations and administers many key committees and subgroups. Through involvement in ISO and IEC, the U.S. has an opportunity to contribute to European Union's standards activities. All comments and proposals made from outside Europe to the European standardizing bodies of CEN and CENELEC pass through the respective national bodies of the ISO and IEC, which in the U.S. is ANSI.

♦ *providing information on standards throughout the world*
As the primary source of standards information in the U.S., ANSI maintains copies of American national standards and drafts, ISO and IEC standards, and drafts, proposals of regional groups associated with the European Union (such as CEN and CENELEC), and the specifications of 90 national standards organizations that belong to the ISO (see Appendix). ANSI promotes the adoption and recognition of these standards and keeps the public informed through a variety of publications.

ANSI Member Participation. ANSI participation is open to all persons who might reasonably be expected to be, or who indicate that they are, directly and materially affected by the standard activity. Participation is not conditional upon membership in any organization and is not unreasonably restricted on the basis of technical qualification or other similar requirements.

Groups that are generally interested in standards development include:

- product producers
- users
- consumers/consumer organizations
- members of the public who are directly affected
- distributors and retailers
- government (users, general interest)
- industrial/commercial

- insurance
- labor
- manufacturers
- professional societies
- regulatory agencies
- testing laboratories
- trade associations
- technical experts

Notification of new standards, revisions, and establishment of new consensus-developing groups and canvass groups is provided by ANSI. Membership on ANSI committees is intended to occur without dominance by any single interest. Membership policy requires that:

- no single interest constitutes more than one-third of the membership of a committee dealing with safety
- no single interest constitutes a majority of the membership of a committee dealing with product standards

Examples of the distribution of membership on the ANSI/HFES 100 (see Chapter 5) and ANSI Z 365 (see Chapter 6) ergonomic standards committees are shown in Table 1–10.

The views and objectives of all the members of the standards development committee (participants) are to be considered during standards development. Members are to be advised of the disposition of, and reasons for, the committee objections. Unresolved objections and substantive changes made in a proposed standard are to be reported to the consensus-developing group or canvass list in order to provide all with an opportunity to respond, reaffirm, or change their votes. Consensus is reached when all views and objections have been considered and a concerted effort has been made toward their resolution.

Role of ANSI Members. ANSI members help to govern ANSI by determining policies, procedures, and long-range plans for stan-

Table 1–10 Membership distribution of ANSI ergonomic standards committees

Representation	ANSI / HFES 100	ANSI Z 365
Government	1	6
Industrial/ commercial	23	21
Insurance	1	3
Labor	1	7
Technical experts, consultants	20	10
Research centers and universities	2	13

dards development. They consider commercial and industrial needs for standards, both nationally and internationally. Based on these needs, members encourage ANSI to initiate new activities for standards development, accelerate current work, and promote the review and improvement of existing standards.

Role of ANSI Standards Developers. ANSI makes it possible for the U.S. to participate in international standardization activities. This participation is a key factor in global marketing. The relationship of ANSI in the global standards arena is shown in Figure 1–2. ANSI is a member of the ISO and the IEC via the U.S. National Committee.

In 1990, ANSI established a program to synchronize the review and approval of its national standards with international standards. The goal of this program is to ensure that ANSI standards are identical to international standards, thus optimizing consistency. Standards written at the international level can be concurrently processed at the national level.

Following international approval, an international standard may be submitted to ANSI for approval by any ANSI-accredited standards developer. Developers of American national standards

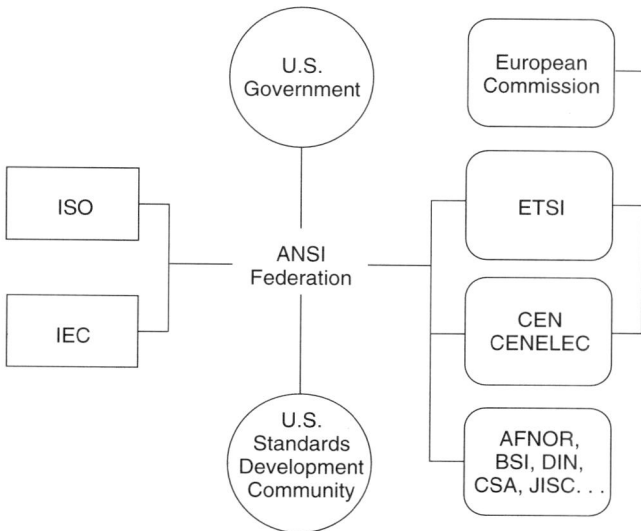

Figure 1–2 Relationship of ANSI to global standards organizations

are required to consider international standards as a basis for the national standard being developed. Deviations from this policy may occur for the following reasons:

- national security requirements
- the prevention of deceptive practices
- the protection of human health or safety, the health of animal or plant life, or the environment
- fundamental technological problems

Standard developers from national trade, technical, professional, consumer, and labor organizations voluntarily submit standards to ANSI for recognition as national standards. Many of these organizations are members of ANSI and are represented on the boards, councils, and committees that help govern ANSI and that coordinate national and international standardization activities. Standards development committees consist of organizations (preferably national in scope), companies, government agencies, and individuals having a direct and material interest in the activities of the committee. These committees are responsible for:

- developing proposed standards within their scope
- voting on proposed standards within their scope
- keeping the standards they develop up to date
- responding to requests for interpretations of the standard(s) they develop
- adopting procedures and revisions
- considering and acting on proposals for terminating procedures
- acting on other matters as provided in their procedures
- maintaining international consistency (as mentioned earlier)

Committees may also obtain assistance from observers and experts in the development process. Committees form subcommittees (SCs) to work on specific topics. For example, the *ANSI HFES/100* revision committee was divided into four subcommittees to address four different but interrelated topics: computer displays, input devices, furniture and environment (see Chapter 5). All standards developed in a subcommittee must be approved by the general committee before distribution for public vote (canvass).

Standards Approval—The Canvass. After an ANSI approved committee produces a draft standard, it submits the document to ANSI which publicly announces and distributes it to individuals and companies who have asked to be on the canvass list. The canvass list is developed by the standards sponsor and approved by ANSI. Any interested party with a direct and material interest has a right to express a viewpoint regarding a new/revised standard, and if dissatisfied, can appeal during this process. The canvass participants can submit comments on a draft proposal within three months of its publication. The standards committee must then review and respond to each comment within four weeks of the canvass deadline. Standards are supposed to be published and made available to the public as soon as possible but no longer than six months after they are completed and receive canvass approval (see Table 1–11). Standards supplements and addenda must be reviewed at least every five years to determine if they should be reaffirmed, revised, or withdrawn.

Relationship between ANSI and the U.S. Government. ANSI has a cooperative relationship with all levels of the U.S. government. Federal agencies and state and local authorities are ANSI members, and their representatives serve on ANSI boards and councils. Coordinating committees, formed with the Occupational Safety and Health Administration (OSHA) (see Chapter 6) and Consumer Product Safety Commission (CPSC), provide forums to discuss activities affecting the voluntary and governmental standards communities.

Table 1–11 ANSI standard approval process

	Months									
	1	2	3	4	5	6	7	8	9	10
Canvass	■	■								
Response to canvass				■						
Publication of final standard					■	■	■	■	■	■

In addition, ANSI's advice is sought by federal agencies, congressional committees, and state and local legislative bodies. An increasing number of ANSI-approved standards have been adopted or referenced by government agencies on issues of health, safety and public welfare, procurement and cost reduction, and public services, such as transportation and communication.

U.S. Input to ISO. ANSI authorizes Technical Advisory Groups (TAGs) to formulate U.S. official positions for ISO technical committees and subcommittees and to participate in ISO technical activities, and operate in compliance with ANSI policy. TAG activities are coordinated by administrators appointed by ANSI to be responsible for ensuring procedural compliance. TAG's responsibilities include the following:

1. Recommend registration, or change in status, of ANSI as a P (participating) or O (observer) member on an ISO technical committee or subcommittee
2. Initiate and approve U.S. proposals for new work items for consideration by an ISO technical committee or subcommittee
3. Initiate and approve U.S. working drafts for submittal to ISO technical committees or subcommittees (and to appropriate working groups) for consideration as committee drafts
4. Determine the U.S. position on an ISO draft international standard, draft technical reports, committee drafts, ISO questionnaires, draft reports, or meetings, etc.
5. Provide adequate U.S. representation to ISO technical committee or subcommittee meetings, designate heads of delegations and members of delegations, and ensure compliance with ANSI's delegates policy
6. Determine U.S. positions on agenda items of ISO technical committee or subcommittee meetings and advise the U.S. delegation of any flexibility it may have on certain issues
7. Nominate U.S. technical experts to serve on ISO working groups
8. Provide assistance to U.S. secretariats of ISO technical com-

mittees or subcommittees upon request, including resolving comments on drafts of international standards, draft technical reports, and committee drafts

9. Identify and establish close liaison with other U.S. technical advisory groups in related fields, or identify ISO or IEC activities which may overlap the TAG's scope
10. Recommend to ANSI the acceptance of secretariats for ISO technical committees or subcommittees
11. Recommend that ANSI invite ISO technical committees or subcommittees to meet in the U.S.
12. Recommend to ANSI the U.S. candidates for the chair of ISO technical committees or subcommittees and U.S. convenors of ISO working groups

ANSI VDT Ergonomic Standard. The Human Factors and Ergonomics Society (HFES) is the ANSI authorized organization responsible for the development of ergonomic standards for computers and related equipment and environments. The HFES developed the ANSI standard for visual display terminal workstations, associated furniture, and work environments (ANSI/HFS 100: *Human Factors Engineering of Visual Display Workstations*, 1988, see Chapter 5). The ANSI/HFS 100 standard was used as the basis for many parts of the ISO 9241 standard. The proposed revised version of this standard includes specifications for keyboards, non-keyboard input devices, displays, furniture, and environmental factors. The ANSI standard for human computer interaction is ANSI HFES 200 (see Chapter 5).

ITI

The Information Technology Industry Council (ITI)—formerly known as the Computer and Business Equipment Manufacturers Association (CBEMA)—represents the interests of leading U.S. companies providing information technology services. Its members include computer, business equipment, and telecommunications hardware and software companies. The major functions of ITT are to:

1. Develop and advocate of public policies beneficial to the information technology industry

2. Participate in all pertinent standards programs worldwide
3. Conduct ongoing industry councils to improve the business operations of its members
4. Provide a forum for executives to work issues across national borders

ITI members are active participants in ISO and ANSI standard development and make contributions to ISO ergonomic standard development through ANSI. Most of the ANSI representatives on the ISO 9241 development committees are from ITI member companies. In addition, many representatives of ITI member companies are active participants in ANSI ergonomic standard development activities.

Summary

Ergonomic standards include specifications that have local, national, and international influence and ramifications. They affect the design of a variety of computer products, including hardware, software, and their associated equipment and work environments. Besides being critical to the computer industry, ergonomic standards are also important in such industries as transportation, telecommunication, and nuclear power.

Ergonomic standards and guidelines exist in most major high-technology marketing countries of the world. ISO creates ergonomic standards for anthropometry, system interaction, and environments. ISO ergonomic standards are based on input from national standard bodies. The U.S. standard agency is ANSI which represents the U.S. in ISO. HFES is the ANSI sanctioned organization authorized to develop ergonomic standards for computer hardware, software, and associated furniture and environments.

References

1. ANSI. Procedures for synchronization of the national and international standards review and approval process.
2. ANSI. Procedures for U.S. participation in the international standards activities of the ISO.

1972 – present

Chapter 2

History of Ergonomic Computer Standards

Overview

This chapter describes the emergence of ergonomic computer workstation and environment standards and other types of requirements. The activities of key contributors to the development of ergonomic standards are also discussed. Standards and legislative ergonomic activities in the United States are presented as well as the role labor plays in developing ergonomic standards.

Emergence of Ergonomic Standards

The Role of Sweden and Germany

Ergonomic standards for computers began in Northern Europe because of concern there for worker health and safety associated with the use of the visual display terminal (VDT). Sweden was the first country to develop a national VDT work ordinance for computers in 1979.

In 1972, the German standards organization (DIN) commissioned the Technical University of Berlin (TUB) to study VDT workplaces and make recommendations regarding their design. The report generated from this study became the basis for a book of VDT guidelines (*Visual Display Terminals*) co-produced by German and British authors in 1980. This book was the first international publication that provided design and use specifications for VDTs, their associated furniture, equipment, and environmental factors on user health, safety, and work organization. Its contents included not only specifications for CRT images, keyboards, radiation emissions, and data transmission but provided information on vision, anthropometry, and mental load to explain the necessity for the specifications. This book became a basis for multinational ergonomic requirements for the next decade.

As a result of the TUB study, the book *Visual Display Terminals,* and increasing concerns regarding the safety and health of VDT workers, Germany issued *Safety Regulations for Office Working Places* (ZH 1/535) in 1976 and a *Regulation for Display Work Places in the Office Sector* (ZH 1/618) in 1980 (see Chapter 4). ZH 1/618 included requirements for:

- visual display units
- microfilm readers
- keyboards
- documents and document holders
- VDT tables
- chairs and footrests
- equipment arrangement
- space requirements
- illumination
- ambient climate
- visual exams
- user training

ZH 1/618 was the first ordinance to specify measurement techniques for assessing the quality of ergonomic features of VDTs and their associated equipment.

Shortly after ZH 1/618 was instituted, Germany created a multi-part hardware and software standard: DIN 66234 *VDT Work Stations*. This standard initiated computer ergonomic standardization in other European countries (see Figure 2–1) and was used as a basis for several of their standards. During the 1980's,

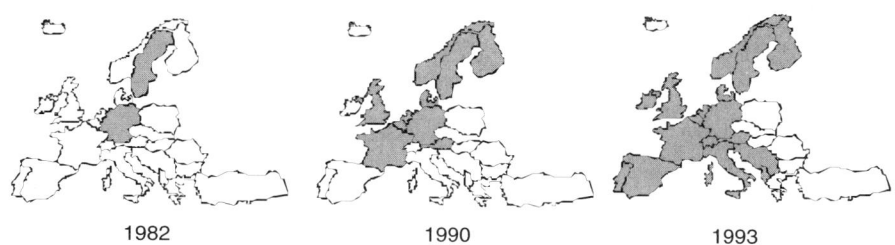

Figure 2–1 Spread of ergonomic standards in Europe from 1980 to 1993 (indicated by shaded areas)

ZH 1/618 and DIN 66234 also became references for the development of ergonomic standards in countries outside Europe.

Other European Countries

While Germany was conducting the TUB study, France instituted a law requiring medical examinations of VDT users. Two years later, France started safety and health inspections of VDT workplaces. The first VDT ergonomic guidelines—*Ergonomic Aspects of Visual Display Terminal*—were published by the British Health and Safety Executive in 1983. By 1990, half of Europe had some type of national standard, ordinance, or guideline assessing the design and use of VDTs and their workplaces.

Beyond Europe

The wave of VDT requirements in Europe soon spread to other areas of the world. In 1984, the first VDT guidelines were published in Australia. In Japan, the first ergonomic display standard was created in 1984 (see Chapter 7). Canada instituted a VDT ergonomic standard in 1989 (see Chapter 5).

European Ergonomic Conferences

In the same year that Germany instituted its national safety VDT workplace ordinance (1980), the first international conference on VDT ergonomics, health and safety—"Ergonomic Aspects of Visual Display Terminals"—was convened in Milan, Italy. The second conference on VDT ergonomics was held in Turin, Italy, four years later. The expanding international interest in VDTs and worker health was evidenced by the increase in the number of papers presented from the Milan conference (40 papers) to the Turin conference (76 papers).

During the next few years, sufficient concern for VDT health and safety issues surfaced that the Swedish government hosted the third major European ergonomic VDT conference—"Work with Display Units"—in 1986. The interest in VDT ergonomics had substantially increased since the Turin conference. There were a total of 316 papers presented at the Swedish conference. The expansion of interest in the impact of VDTs on the worker was evidenced by the additional topics. The majority of papers presented at the Milan and Turin conferences addressed vision problems associated with computer terminal use. Although conference papers continued to address vision issues, there was a significant increase in technical papers on biomechanical problems, mental stress, and radiation emission research (see Figure 2-2). It was ironic that the week before the Swedish conference, the nuclear power plant in Chernobyl blew up. Although the levels of radiation emission from display screens were minuscule compared to those emitted from Chernobyl, the concerns voiced at the conference focused on the number of people yearly exposed to VDT emissions, which far surpassed those that were under the Chernobyl radiation cloud. The concerns of radiation from VDTs continue today, although adverse effects from radiation have not yet been proven.

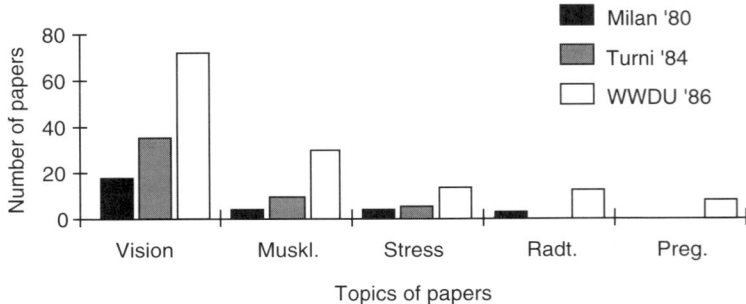

Figure 2–2 Number and topic of health and safety papers presented at early VDT conferences in Europe

International Spread of VDT Ergonomic Issues

Thus between 1973 and 1990, several major events that influenced the national establishment of ergonomic standards occurred around the world. The events included the creation of standards, ordinances, and guidelines, the occurrence of VDT conferences, inspection audits, and the publications of technical papers (see Table 2–1).

Table 2–1 Major events resulting in the spread of ergonomic VDT standards and legislation

Date	Location	Event
1972	Germany	establishes VDT standards committee
1975	France	creates noise level decree for office products
	Belgium	creates product refusal for ergonomic reasons
1976	Germany	institutes safety regulation for office workplaces
1977	France	creates law for medical checkups for VDT users
	Finland	creates safety & health guidelines for VDT displays and furniture
1978	Germany	publishes University of Berlin (TUB) report on VDT workplaces
1979	Sweden	creates first national directive
	France	initiates inspection of VDT workplaces
1980	Great Britain	publishes Health & Safety Executive VDT guidelines
	Germany	institutes safety regulation for VDT workplaces
	Italy	hosts the first ergonomic VDT conference
1981	U.S.	(Department of Health) issues statement on VDT health
		National Academy of Sciences VDT conference
	Austria	initiates shift work laws for VDT use
	Netherlands	creates directives for use of VDTs
1982	Germany	starts compliance testing of VDTs
1983	Great Britain	publishes VDT Health and Safety Executive guidelines and initiates creation of ISO VDT standard
	U.S.	initiates ANSI/ HFS VDT standard
	Austria	creates standard for VDT workplaces

(continued)

Table 2–1 *Continued*

Date	Location	Event
	Norway	proposes regulation for work at VDTs
1984	Italy	hosts the second ergonomic VDT conference
	Australia	publishes VDT guidelines
	Japan	publishes VDT display standard
1986	Sweden	hosts the third VDT (first WWDU) ergonomic conference
	Sweden	creates directives for reading/scanning of VDTs
	Italy	hosts the fourth ergonomic VDT conference
1987	Europe	publishes WHO report on VDTs and worker health
	Sweden	establishes MPR radiation measurement requirement
1988	United States	publishes standards on VDT workstations
1989	Canada	publishes VDT office standard

World Health Organization VDT Report

One year after the Swedish "Work with Display Units" conference (1986), the World Health Organization (WHO) published a report on VDT worker health. This 1987 report was prepared by the Swedish National Board of Occupational Health and Safety. It was the most comprehensive international report on health issues at that time and contained a summary of research on visual, musculoskeletal, stress, skin, and reproductive disorders and concerns associated with VDT work. It specified guidelines for working with VDTs, provided answers to common questions regarding VDT health issues, and provided recommendations for research. It became the basis for the first international directive (the EU *Display Screen Directive*) on VDTs (see Chapter 4).

The research papers presented at the Italian and the Swedish conferences on *Work with Display Units* and the WHO report had a major impact on international and national standard development and international labor union activities.

The United States Situation

In the United States, there was about a ten-year lag in initiation of VDT ergonomic standardization. In addition, the standardization focus in the U.S. was (and still is) on design effectiveness and user performance rather than on safety and health issues emphasized in European standards. The reasons for this difference lie in the evolution of ergonomics in the U.S. and the focus of the U.S. on worker efficiency.

Ergonomic standards in the U.S. have traditionally been composed of quantitative design specifications; ergonomic analysis has traditionally focused on objective measures of human performance. Standards and ergonomic analysis in several other countries traditionally focused on worker health, safety, and comfort; their standards specify and measure physical features, but they use both performance and comfort metrics to assess product compliance to standards. They thus view product compliance measurement from a systemic rather than a component process. Interestingly, although the U.S. historically focused on evaluating worker performance, its standard organizations have traditionally focused on the specifications of equipment design and measurement through instruments rather than measures of user performance.

The first major national U.S. event addressing VDT worker health and safety occurred almost ten years after the German TUB study began. In 1981, the National Academy of Science sponsored a conference on VDTs in Washington, DC. That same year, NIOSH requested the NAS to review studies of visual issues associated with use of VDTs. In 1983, the committee appointed by the NAS published their findings in *Video Displays, Work and Vision*. All but one member of the committee concluded that using a VDT was not associated with health problems. That same year, the U.S. initiated development of a VDT workstation standard which was published in 1988. Since then, the U.S. has passed two federal regulations on office equipment design and use (see Chapter 5).

The Ergonomic Standards Debate

 Hardware Issues

The intent of developers of ergonomic standards is that standards be based on quantifiable scientific research. However, the best intentions have not always resulted in scientifically substantiated specifications. Some specifications of the early VDT ergonomic standards were based on qualitative data and judgments of well intended individuals who were criticized for not having the appropriate knowledge and expertise to specify certain technical requirements. This opinion was widely held in the United States, particularly by the U.S. computer industry, which actively and publicly criticized several ergonomic standards requirements.

An example of a standards requirement that caused a considerable number of complaints from the United States was the requirement for positive display imaging (that is, white background screens with black text). This requirement first occurred in Germany because of the desire to minimize visual stress supposedly caused by excessive contrast between a dark background screen and white documents. The standard was also intended to reduce the visual stress resulting from the perception of reflections on dark background screens.

Later analysis and research revealed that there are many factors that contribute to visual comfort and performance and that the human eye is not excessively or dysfunctionally stressed by viewing dark background screens when alternatively looking at white paper in offices with ambient illumination of 500 lux. Research also indicated that white background screens were often a source of visual discomfort because there is more light emitted from a white background CRT screen than from the white characters on a dark background screen. The greater light emitted from white screens often becomes a source of glare, especially in offices with low ambient light levels (such as 200 lux)—as is often the case in computer aided design work areas).

Ergonomic studies also showed that screen flicker was more

Figure 2–3 Appearance of negative (a) and positive (b) polarity characters

easily perceived on white background screens and was a major source of visual annoyance and complaints. In addition, the amount of brightness of white background screens caused the lines of black alphanumeric characters to appear thinner than those of the white characters on a black screen (see Figure 2–3). Although this enhanced the appearance of edge sharpness (resolution), black characters were not as easily visible, particularly at viewing distances further than 20 inches and for users with visual problems (such as low acuity and cataracts). Ergonomic research has also shown that the colors of commonly used thin fonts were more difficult to identify on a white background screen than on a dark gray or black background.

Other standard requirements that have since been shown to be invalid, unnecessary, or harmful (causing biomechanical or physiological problems) include requirements for palm rests, four-legged chairs, and eye-height screens. Recent research has shown that unpadded palm rests can cause wrist disorders; three-legged office chairs can be as resistant to tipping as four-legged chairs if the diameter of the base is sufficient; screens at eye height cause significantly more drying of the surface (cornea) of the eye and more visual and neck discomfort than screens placed below the eye.

Subsequent findings of ergonomic research led to the modification of some of the standards and the incorporation of appropriate requirements into many new standards. However, the effects of many of the older and inappropriate requirements (such as unpadded palm rests, four-legged office chairs, and white background screens) continue to pervade many ergonomic standards, brochures, textbooks, training courses, and consultant recommendations.

Cost Concerns

In addition to the concern of the scientific validity and measurement methods of early ergonomic standards, the U.S. computer industry was concerned about the cost of redevelopment, fabrication, and testing associated with complying to the ergonomic standards of the early 1980s. For example, the hardware requirement first established by Germany in 1980 for detached and low profile keyboards resulted in a furor of complaints and excuses from the U.S. computer industry about why they could not economically produce them. When Germany would not withdraw its keyboard requirement, the keyboards became commonly available and inexpensive within a few years. The cost of a low profile, detached keyboard has dropped from over $100 in the early 1990s to as low as $16.

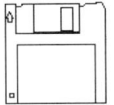
Software Issues

In 1986, Germany released its first ergonomic standard on software user interface (Part 8 of DIN 66234, see Chapter 4). Once again, a major objection was raised by the United States. The U.S. felt that the standard was inappropriate because its requirements had not been scientifically validated and there were no explicit criteria against which software could be tested to determine compliance.

German publications, on the other hand, indicated that the contents of their new standard was justified. The creators of DIN 66234 believed that software ergonomic standards were different from technical standards: *technical standards* were strictly dependent on *dynamic* state-of-the-art technologies, while *egonomic*

The Ergonomic Standards Debate

Table 2–2 German view of technical vs. ergonomic standards

Standard characteristic	Technical standard	Ergonomic standard
State	dynamic	static
Aims	cost minimization	human efficiency, safety, health, and comfort
Basis	compromise and consensus	facts
Focus	product centered	human centered
Importance	prescribed in the standard	left to designer judgment
Target	average users	specific users

standards should be based on human factors engineering principles, which are *constant* over time. In addition, they felt that a major aim of technical standards was to reduce cost and that the goal of ergonomic standards was to optimize user efficiency, health, and comfort. The Germans felt that although compromises were acceptable in creating technical standards, ergonomic standards concerning health and safety principles and specifications should not be compromised. The German viewpoint comparing technical and an ergonomic standards is summarized in Table 2–2.

In addition, the intention of Germany was not to create a *product-centered* standard, but to create a standard focusing on *user interaction* (including problems and task requirements). This point is obvious in the following statement from Part 8 of the DIN 66234 standard, which contains the type of qualitative phrasing of specifications and measurements that caused the U.S. dissatisfaction:

> . . . the dialog cannot be designed without reference to the task and the organization of work . . . if a test is to be made as to whether the dialogue requirements formulated in this standard have been met, this will also generally be carried out with reference to a specific task. Inspection criteria may be derived from an inspection of the task and its performance which are needed for the verification of the characteristics of a dialogue system.

Incidents like these demonstrated the basic difference in standard philosophy held in Germany and the United States and characterized the historical differences between the ergonomic focus in Europe and human factors engineering in the United States. Although Germany may appear to be a regimented and standards-based country, its intent is to produce standards sufficiently flexible to inspire improved and novel designs as well as educate designers regarding requirements. German publications present the belief that software ergonomic standards should be formulated as guidelines rather than mandatory requirements and should not be mandatory unless supported by agreed-to scientific evidence. This explains this kind of phrasing in Part 8:

> With practice in the use of a particular dialogue system, the user becomes accustomed to its response time and thus develops certain expectations in this respect. Procedures of the same type, particularly those having short processing times, shall have similar response times. If a calculable response time is going to deviate considerably from the normal response time, then the user shall be informed of this.

Evidently, this philosophy is acceptable to a majority of ISO member countries, because, as a result of national consensus voting, the German philosophy and basic requirements known as "primitives" have been incorporated into the software parts of ISO 9241 (see Chapter 3). These primitives include self-descriptiveness, controllability, suitability for the task, conformity with user expectations, and error tolerance and have been incorporated as explicit statements in several parts of ISO 9241.

When Germany first released a draft of its software standard DIN 66234, Part 8, a group of U.S. human factors engineers from ITI (then CBEMA, see Chapter 1) met to compose a critique of Part 8 to be delivered to DIN. Although their comments did not result in any substantial changes to the German standard, the U.S. group soon became active in national and international standard development. It became organized into the Human Computer Interaction committee sanctioned by the Human Factors and Ergonomics Society and the American National Standards Institute. The original intention of the group was not to create a standard, but to advise the Human Factors and Ergonomics Society on software ergonomic activities world-wide and to report on the feasibility of producing HCI standards. However, when the committee realized that HCI standards were going to be adopted in the U.S. and other countries with or without human factors engineering input, they decided to generate alternative standards proposals, based on stringent criteria for guidelines and standards. Most of their input has since become integrated into the software parts (10–17) of ISO 9241. The committee is now evolving their collection of documents and human factors into a software HCI standard for the U.S. (see Chapter 5).

 Industry Concerns Regarding Consistency

In addition to the concern regarding standard validity in the early 1980s was the concern regarding lack of consistency of standards and national guidelines.

Until the 1980s, most ergonomic specifications in the U.S. were recommendations from reference books, scientific journals, or corporations. For example in 1975, IBM published a set of guidelines for *Man/Display Interfaces.* The same year, Bolt, Berank, and Newman published *Display Specifications Procedures* for the Data Systems Divisions of the U.S. Department of Agriculture. It included specifications for system response time, menu selection, format, and errors messages.

As with the VDT hardware requirements of the 1980s, software specifications relating to VDTs began to be consolidated into national and international standards, regulations, guidelines, and marketing requirements. One of the advantages of the movement toward the internationalization of software specifications into ergonomic standards was the consolidation of previous textbook ergonomic principles and practices into a single consistent source.

Different specifications occurred across as well as within countries and organizations. For example, the German DIN 66234 standard specified a minimum display character height of 2.6 mm; the Swedish TCO labor guidelines specified a minimum

Table 2–3 Examples of ergonomic requirement inconsistencies

Feature	United States	Germany	Great Britain	Sweden
DISPLAYS				
Design viewing distance	400 mm	≥ 500 mm	350–600 mm	600 ± 100 mm
Character size	≥ 18 arc min.- low definition characters; 15 arc min.- high definition characters.	2.6 mm / 18 arc min.	≥ 16 arc min.	4 mm (20 arc min. at 600 mm)
Character format	≥ 5×7 7×9 preferred	≥ 5×7	≥ 5×7: upper case & numbers; ≥ 7×9: lower case	14 × 11 for full width capital letter
Interline separation	≥1/2 character height	one dot position/ 10% character height	[no requirement]	≥ 70% character height
Refresh rate	[no requirement]	flicker free	flicker free	≥ 50 Hz
KEYBOARDS				
Thickness	[no requirement]	≥ 30 mm at home row	< 70–90% of upper & lower arm	[no requirem] [no requirem]
Slope	17–18°	[no requirement]	10–15°	

character height of 4.0 mm. These, and other, differences (see Table 2–3) were considered to be a significant problem for the computer industry because of the high cost of producing VDTs with different design features for different countries and organizations. As a result, the computer industry became very actively involved in a dual role: while computer companies and their trade associations (such as CBEMA and ECMA) lobbied against what they believed to be "pre-mature" and insupportable standards and laws, they also assisted standards organizations in creating standards that they believed were scientifically valid and accurately and quantitatively measurable.

 ## Formation of ISO Ergonomic Standards

The ISO technical committee responsible for ergonomics (TC 159) was established in 1975 to address many of these concerns and problems with ergonomic standards. In 1983, the British Standards Institution addressed these by creating an initiative to develop a multi-part, international (ISO) standard based on internationally recognized ergonomic principles and research findings. A committee of ergonomists was convened in Bournmoth, England to create a standard for visual display workstations. This standard was to be a performance, safety and health based standard. Although software user interface design specifications were not originally included its scope, they were soon added in 1987. The result was a seventeen part ISO standard consisting of hardware specifications for computer displays, input devices, furniture, and environmental factors as well as specifications for software user interfaces (see Chapter 3).

 ## Europe's Solution

To address the problem of national inconsistency in ergonomic computer standards, European countries formed Agreements (see Chapter 4), which required them to exchange technical information and coordinate standardization. One example was the Lisbon Agreement on technical information exchange between ISO and CEN. The three primary purposes of the Lisbon Agreement were to:

- ensure the flow of information between ISO/TCs/SCs and CEN/TCs/SCs
- to avoid duplication of work between these technical bodies
- to increase the transparency of their activities

Another example was the Vienna Agreement for technical cooperation between ISO and CEN. These agreements were supported by the Commission of the European Communities (the administrative and executive arm of the European Union) in the interest of international trade and development of common European standards that would be consistent with international standards. The cooperation of ISO and IEC with CEN and CENELEC were intended to bring the larger international community into the process of development of needed standards and to enhance coordination with the European market.

 **Ergonomic Legislation in the United States**

In the early 1980s, the United States Congress and Senate began conducting national hearings on VDT health and safety. At the same time, ordinances regarding VDT design and use were being introduced into state legislation. Between 1980 and 1986, over 200 bills on VDT design and use were introduced in 35 states (see Figure 2–4). The computer industry and business establishments successfully lobbied against all the bills. As a result, almost all of the bills were defeated.

These bills included requirements on:

- VDT equipment
- environment
- work activities
- vision
- testing and maintenance
- technology refusal
- upgrade notification

The VDT equipment requirements included adjustability, treatment for screen glare and radiation shielding (see Table 2–4). The bills addressing the work environment included specifications for chemical, heat, noise, and light emissions and control. Work activities specifications addressed work limits, keystroke monitoring, and considerations for pregnant workers. The vision specifications required periodic eye exams and provision of VDT viewing glasses. Some of the state bills required the transfer of pregnant workers to jobs other than those using VDTs; other bills

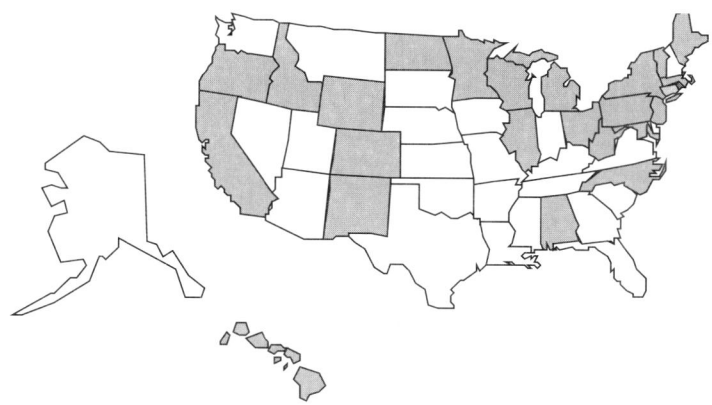

Figure 2–4 States introducing VDT legislation between 1980–1984 (indicated by shaded areas)

allowed workers concerned about health effects of VDTs to transfer to other jobs.

Several U.S. states have attempted to convert the 1988 ANSI ergonomic VDT workstation standard (see Chapter 5) into law; none have yet been successful. This type of legislation would change the ANSI standard from being voluntary to mandatory.

Table 2–4 State legislation proposed on VDT equipment (1980–1984)

	VDT Equipment Issue		
State	**Adjustability**	**Screen Glare Treatment**	**Radiation Shielding**
California	✓	✓	✓
Illinois	✓	✓	
Maine	✓		
Massachusetts	✓	✓	✓
Minnesota	✓		✓
New York	✓	✓	✓
Ohio	✓	✓	✓
Oregon	✓	✓	✓
Pennsylvania	✓	✓	✓
Rhode Island	✓	✓	✓
Wisconsin	✓	✓	

Products that did not comply with its requirements would not be considered for purchase by state agencies and, like the EU *Display Screen Directive* (see Chapter 4), after a specific period old products would have to be replaced with complying products.

CTD Laws

A recent effort in state legislation is the California Occupational Safety and Health Standards Board proposal of the regulation *Prevention of Cumulative Trauma Disorders (Ergonomics)*. It is the result of a study completed by CAL-OSHA in 1987, conclusions of a committee to study CTDs at the workplace and research presented at major international scientific conferences. The purpose of the CAL-OSHA regulation is to establish minimum requirements for controlling exposure risk of developing cumulative trauma disorders (CTDs). The regulation applies to all types of repetitive work, including VDT use and specified procedures for gathering information on cumulative trauma disorder (CTD) risks, worksite evaluations, control measures, medical management, and training.

U.S. Laws for the Disabled

The only national ergonomic laws that currently exist in the United States are for the disabled. They are the *Americans with Disabilities Act* (ADA) and the General Service Administration (GSA) ordinance for handicap use (see Chapter 6). The Americans with Disabilities Act was created by the U.S. Congress to provide disabled citizens greater employment opportunities and public access. Although most of the requirements in the ADA deal with public access, hiring practices, and business conduct, the ADA also includes specifications for workplaces, including those where VDTs are used. The General Services Administration (GSA) ordinance is more directly related to VDTs, because it requires that handicapped individuals be provided with easily usable electronic office equipment. This law was a result of a mandate by the U.S. Congress that "guidelines for electronic equipment accessibility by the handicapped be established and adopted and that government agencies comply with these guidelines with respect to electronic equipment, whether purchased or leased."

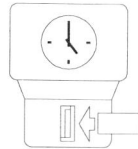

Labor Union Activities

Although VDT ergonomic standards have been traditionally created by representatives from national standards organizations, labor unions have played a major role in the consideration and implementation of these standards. As blue collar workers moved into white collar offices and jobs, the tools and hazards of their work changed. Membership in unions that formerly protected workers from dangerous equipment and harsh environments did not seem necessary for office computer work. However, labor unions were motivated to retain their membership and so they became actively involved in identifying potential safety, health, and social problems in the VDT office environment and in offering protection policies for VDT workers.

Labor unions thus played a significant role in raising and spreading concerns regarding VDT health and safety, encouraging standards, introducing regulations, motivating the creation of guidelines, and using ergonomics for arbitration negotiations. For example, in February 1985 FIET, an international white collar union, created guidelines on VDTs. At the same time, the United Kingdom's Health and Safety Executive and West Germany's Industrial Injuries Institute drew up codes of good employer practice toward VDT workers. In 1986, the Swedish Central Organization of Salaried Employees (TCO) created a set of VDT guidelines (*Screen Checker*) that continues to be the most stringent requirements for VDTs in the world (see Chapter 4).

However, as with ergonomic standards, union guidelines exhibited inconsistencies (see Table 2–5) and the validity of their requirements were considered suspect by industry and many scientists.

Table 2–5 Labor union guidelines and inconsistencies

Display Feature	U.S. APEX* 1980	U.S. NYCOSH* 1980	Australia ACTU-VDHC* 1982	Sweden TCO* 1986
Character size	1/200 viewing distance	3/16 viewing distance	—	3.8 to 4.5 mm
Character format	5×7/7×9 minimum	—	7×9 minimum 9×14 preferred	—
Interline separation	1 character height	—	1 character height minimum	to allow easy recognition
Character and background color for monochrome screens	light green characters on dark green screen; yellow or white characters on black screen	light green, yellow, or white characters on dark background	yellow characters on dark green; green characters on dark background	black text on white background; green or yellow text on black background

* APEX: Association of Professional, Executive, Clerical and Computer Staff (Trade Union Affiliate)
* NYCOSH: New York Committee for Occupational Safety and Health (AFL/CIO affiliate)
* ACTU-VTHC: Australian Council of Trade Unions and the Victorian Trades Hall Council
* TCO: The Central Organization of Salaried Employees

While labor was encouraging standardization of ergonomic requirements, the computer industry maintained that since there was no absolute proof that VDUs cause health risks, public intervention was not necessary or appropriate. Labor unions believed that there was sufficient cause for concern to disclaim industry's belief that none of the concerns were well founded. The areas upon which labor's doubts were focused were:

- radiation
- pregnancy risks
- eye stress
- psychological stress
- muscular strain and disorders
- posture problems
- face rashes

Labor Union Activities

The union leadership continues to believe the following:

- Symptoms of VDT user disorders, particularly musculoskeletal symptoms, are increasing, especially in individuals who use VDTs extensively and intensively.
- Actions are necessary by vendors and users to minimize likely health hazards.
- Large-scale research on VDT health and safety issues by government and independent centers of competence is necessary.
- Legislation of VDT design and use is necessary to help protect workers until the results of research are available and can be implemented.
- Filing of worker compensation claims for vision problems, musculo-skeletal effects, repetitive strain injuries (RSIs), and stress-based disorders are necessary.
- VDT problems should be addressed with counseling, expert help, and legal representation.

In the last decade, these beliefs have been dominant themes in union newspapers, brochures, conferences, and membership recruitment activities. For example, in 1986 the AFL-CIO Department for Professional Employees sponsored a national conference on issues of technology change which included ergonomics of computer workplaces. Since this sector of the AFL-CIO represents the largest organization of its kind in the world (over two million professional and technical workers), its influence is significantly far-reaching.

Labor believes that industry is more worried about avoiding liability than in avoiding worker injuries and disorders. Labor's perception of industry's attitude continues to motivate unions to introduce and support VDT ergonomic legislation.

Although unions have historically focused on hardware standards and regulation, in 1986 they announced their plans to initiate regulatory activities in software design and use. Unions continue to push for local, state, and national legislation, executive orders, and standards.

Summary

Ergonomic computer standards have evolved from one national standard in Germany to international standards and legislation. Although Europe was and continues to be a leader in standardization of the VDT workplace and ergonomic requirements, other countries are becoming active in supporting and participating in ergonomic standards development. Although there continue to be many unresolved ergonomic issues, today there is more consensus on basic product design and user-interface principles than ever in the history of ergonomics.

Chapter 3
ISO 9241

Overview

This chapter describes the contents of the international ergonomic standard on computer display workstations—ISO 9241: *Ergonomic requirements for office work with visual display terminals*. It presents an overview and examples of the ISO 9241 requirements, its applications, and its parts. It also addresses how compliance with the standard is demonstrated and assessed.

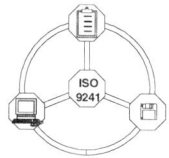 **ISO 9241**

In 1983, the British Standards Institutes initiated development of an ISO performance-based ergonomic standard for computer display terminals and their associated workstations, peripheral equipment, and environments. The purpose of this standard was to promote the health and safety of VDT users and to ensure that they can operate VDT equipment efficiently, safely, and comfortably. The initiative became ISO 9241: *Ergonomic requirements for office work with visual display terminals (VDTs)*. The standard specifies requirements for VDT hardware and software user interfaces and is intended to help designers and manufacturers develop ergonomically appropriate products. The standard is also intended to be used by purchasers who wish to specify requirements for VDT selection and installation or to assess the suitability of existing equipment or environmental conditions.

ISO 9241 also specifies requirements for the VDT workstation, the VDT working environment, and the organization and management of VDT work. It emphasizes the specification of user performance through equipment characteristics. Although ISO 9241 originally focused on hardware requirements for VDTs, it soon evolved to include software requirements.

ISO 9241 is composed of 17 parts addressing four major areas: introduction and general description (Parts 1–2), hardware requirements (Parts 3–9), environmental requirements (Part 6), and software requirements (Parts 10–17) (see Figure 3–1).

Some of the specifications in ISO 9241 are controlled by hardware and some by software (see Table 3–1). For example, the ability of a display to produce a color is dependent on hardware, but the assignment of color to a particular image on the screen is controlled by software. And while it is true that the movement of the screen cursor is controlled by both hardware and software, the size and shape of the cursor is software controlled.

The first two parts of ISO 9241 deal with a general introduction to the standard and guidance on task requirements. Hardware and environmental requirements are contained in Parts 3–9. Software specifications are contained in Parts 10–17.[1]

Part 1: *General introduction* (IS) contains general information about the standard and provides an overview of each of the parts. Some references and definitions necessary for under-

ISO 9241

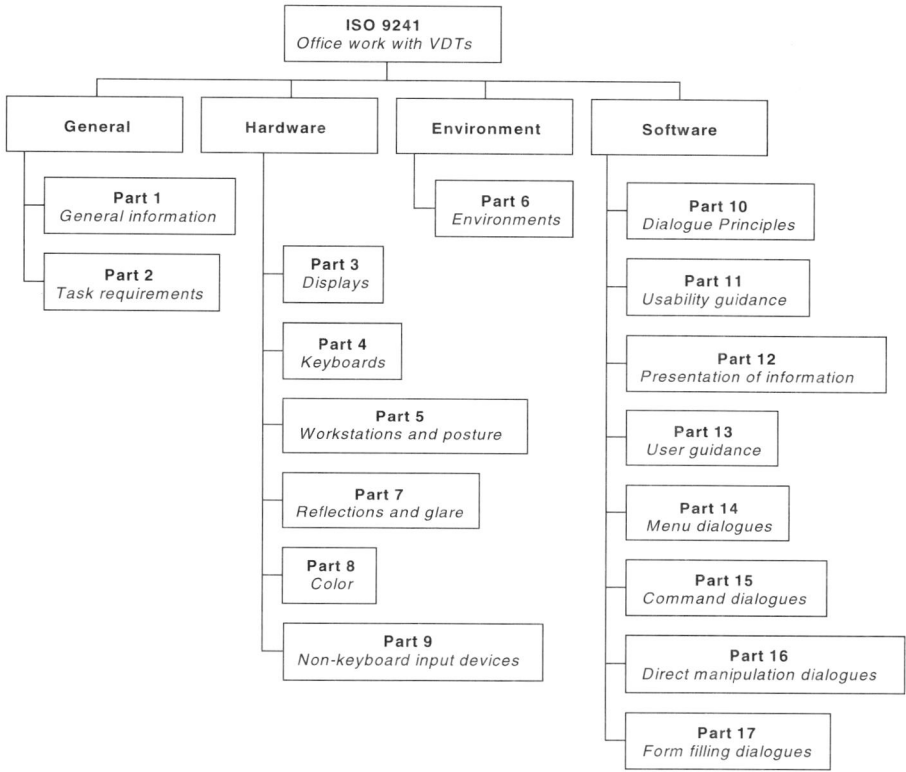

Figure 3–1 Organization of ISO 9241 by subject

standing the standard are also included. In addition, the basis of the user performance approach of the standard is explained. Guidance is given on how to apply the standard.

Part 2: *Task requirements* (IS) discusses the enhancement of user interface efficiency and the well being of users by applying practical ergonomic knowledge to the design of VDT work tasks.

Part 2 specifies that well-designed tasks should:

Table 3–1 ISO 9241 parts and focus

Part No.	Subject	Control Hardware	Control Software
1	General information	✓	✓
2	Task requirements	✓	✓
3	Displays	✓	
4	Keyboards	✓	
5	Workstation and posture	✓	
6	Environment	✓	
7	Reflections and glare	✓	
8	Color	✓	✓
9	Non-keyboard input devices	✓	✓
10	Dialogue principles		✓
11	Usability guidance		✓
12	Presentation of information		✓
13	User guidance		✓
14	Menu dialogues		✓
15	Command dialogues		✓
16	Direct manipulation dialogues		
17	Form-filling dialogues		✓

- facilitate task performance
- safeguard the user's health and safety
- promote individual well-being
- develop user's skills and capabilities

Part 2 also states that well-designed tasks have:

- a variety in activities and skills used
- some degree of individual control over the workspace
- sufficient cohesion so that the task forms an understandable part of the work
- an opportunity for workers to use and enhance their skills and experience
- sufficient feedback that is both meaningful and helpful

Part 2 also provides guidance on how task requirements may be identified and specified within organizations. In addition, it describes how task requirements can be incorporated into the system design and implementation process.

Part 3: *Display requirements* (IS) specifies requirements for visual displays and their images. The user performance objective of this part is that the user should be able to detect and recognize images both accurately and quickly without visual discomfort. The display characteristics that influence detection and recognition are identified. Where appropriate, design guidance on the minimum, maximum, and optimum values of display characteristics are presented. Implications for display characteristics of some of the requirements are shown in Table 3–2.

Table 3–2 Examples of normative requirements and implications in Part 3

Design Feature	Specification	Implication(s)
Viewing distance	≥ 400 mm	All images are to be easily visible at 400 mm
Viewing angle	0–60°	All images must be easily visible down to 60° from horizontal line of sight
Character height	≥ 16 arc min.	For black and white characters only
	≥ 20 arc min.	Colored characters should be larger than black and white characters
Stroke width	1/6–1/12 character height	Some bold fonts may not meet the spec.
Width-height ratio	0.5:1–1:1	Very narrow and wide fonts not acceptable
Luminance modulation	≥ 0.4 (monochrome) ≥0.7 (color)	Noticeable differences intended; same luminance not allowed
Character format	≥ 5×7	Larger characters are required for reading

(*continued*)

Table 3–2 *Continued*

Design Feature	Specification	Implication(s)
Size uniformity	≤ 5%	Same characters cannot vary much in size
Space between characters	≥1 pixel	Less than 1 pixel (stroke width) between sans serif characters and between serifs not acceptable
Space between words	≥1 character width	Less than one typical character ("N") not acceptable
Space between lines	≥1 pixel	Less than 1 pixel not acceptable
Linearity	≤ 2%	Same rows and columns not to differ by more than 5% in length
Luminance	≥ 35 cd/m^2	Minimum brightness and lowest code
Contrast	≥ 3:1	If background is 35 cd/m^2, character must be at least 70 cd/m^2
Luminance coding	≥ 5:1	Brightness codes must be at least 5 times brighter than each other
Flicker	No flicker	Most CRT displays need to be refreshed at least 80 Hz.
Stability	No jitter	Images must not appear to move

The requirements in Part 3 affect both display hardware and software. For example, the hardware components of the cathode ray tube (CRT) produce the pixels that form images on the screen. The contrast between characters and background (see Figure 3–2), is determined by the hardware and software of the display and the sources and direction of light in the environment. There are also interactions between hardware, software, and the environment. For example, increasing the room illuminance reduces the saturation of colors on the screen.

In Part 3, the effect on user performance of display design characteristics can be assessed in the visual display performance

Figure 3–2 Background lightness and visibility

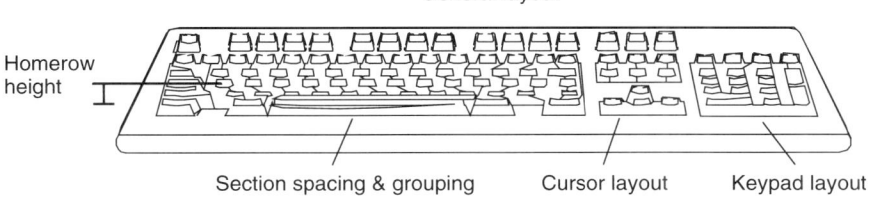

Figure 3–3 Examples of required keyboard features

test (see Chapter 9). User performance is expressed in terms of the minimum accuracy and speed achieved by test subjects in an alphanumeric character detection and recognition test. Comfort is assessed by a rating questionnaire. The test procedures and conditions are specified as well as the required user performance.

Part 4: *Keyboard requirements* (DIS) specifies the characteristics that determine effectiveness in accepting keystrokes from a user. The user performance objective specifies that a user should be able to locate and activate keys accurately and quickly, without discomfort and excessive biomechanical load. Examples of keyboard characteristics that influence keying performance and have specifications in this part are shown in Figure 3–3.

As in Part 3, design guidance and specifications on the minimum, maximum, and optimum values of each characteristic are given. There are several consequences that these specifications have for the design of keyboards (see Table 3–3).

Since keying performance can also be influenced by readability of the labels on the keys, specifications for the design of key labels are also included (see Figure 3–4).

The overall effect of the keyboard design is assessed in a performance test (see Chapter 9). The test procedure and conditions are specified, and the required user performance is specified in

Table 3–3 Examples of keyboard specifications and implications

Design Feature	Specification	Implication(s)
Size	13 mm wide; 19 mm center spacing;	Smaller key sizes on most lap top keyboards will not comply when they are used as office equipment
Shape	Concave	Convex keys are not acceptable
Layout	Any layout acceptable as long as it allows neutral postures	Conventional layouts typically do not result in users keying with hands in neutral postures and thus do not comply
Force	0.5–0.6 N (newtons)	Low displacement required
Labels	5:1 contrast	High lightness and color contrast required

terms of keying accuracy and speed. Keying comfort is determined by a rating scale.

There are a number of other keyboard characteristics that are addressed in a different ISO standard (that is, ISO 9995). This standard includes specifications for:

- general keyboard layout
- alphanumeric labels
- numeric labels
- editing keys
- function keys

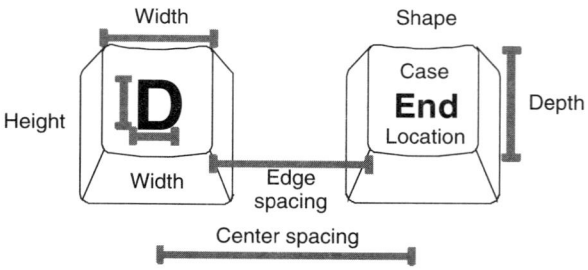

Figure 3–4 Key characteristics specified by Part 4

Part 5: *Workstation requirements* (DIS) specifies the design characteristics of workplaces in which VDTs are used. The performance objective of this part is that the workplace facilitate efficient operation of the VDT and encourage the user to adopt a comfortable and healthy working posture. The characteristics of the workplace that promote a healthy and comfortable posture are identified, and design specifications and guidelines are provided. They require that:

- frequently used equipment controls, displays, and work surfaces are within easy reach
- frequently viewed displays are within easy view
- the opportunity to change position is frequently given
- excessively frequent, repetitive movements with extreme extension or rotation of the limbs or trunk are avoided
- back support, especially for the lumbar region, is present

This standard part explains that applications may exist where the special demands of the tasks and environment may require more stringent requirements to promote efficient operation and healthy and comfortable postures.

Part 6: *Environmental requirements* (CD) specifies characteristics of the working environment in which VDTs are used. It includes requirements for natural and artificial lighting (including glare shielding), sound and noise, mechanical vibrations and electromagnetic fields, and thermal conditions. The characteristics of the working environment that influence efficient operation and user comfort are identified, and design guidelines are presented. Part 6 also includes informative annexes with techniques for measuring:

- direct glare
- noise
- air temperature
- humidity
- air velocity
- thermal radiation

The performance objective of this part is the facilitation of efficient use of the VDT by providing a comfortable and safe work environment. The specifications for this part are more flexible than those in the other parts. It concedes that individuals will vary in their judgments and preferences of acceptable environmental conditions.

Part 7: *Display requirements with reflections* (CD) describe how to maintain usable and acceptable VDT image quality by evaluating the reflection properties of a screen and the image quality of the screen over a range of typical office lighting conditions. It describes the different, but acceptable, limits for reflections on both positive and negative polarity screens.

The specifications of this part are based on office illuminations of approximately 500 lux with minimum and controlled glare sources. They thus apply to illumination conditions in a narrow range and do not apply to very low or very high luminance environments with uncontrolled glare sources or poor luminance balance. Part 7 thus does not apply to offices with low illumination levels like those often used at CAD workstations.

Although this standard part is intended primarily for displays used for reading or data entry tasks, it may also be appropriate for tasks that make greater visual demands (such as graphic and computer aided design), depending on the illumination levels and conditions of the work environment.

Part 8: *Requirements for displayed color* (DIS) states specifications for display color images, color measurement metrics, and visual perception tests. The performance objectives require that users be able to detect, discriminate, and identify colors. However, the color specifications in this part are for images on computer displays viewed by users with normal color vision. Thus displays that conform to this standard part will not be optimized for persons with color vision deficiencies (about 8–10 percent of males and 1 percent females).

The majority of specifications of Part 8 address the following:

- default color palettes
- minimum size for color characters and images
- misconvergence limits
- uniformity limits
- colors for images and their background
- number of colors
- labeling for colors

This part addresses perceptual components of color (such as color detection and recognition) as well as cognitive components of color interpretation (such as identification of a specific color and the number of colors that should be used for information coding). It applies to characteristics of colors assigned to text and simple graphics on word processors as well as to text and simple graphic images on CAD/CAM, and desktop publishing applications, and medical and scientific instrumentation displays.

Since color images are controlled by both hardware and software, the requirements in this standard part are relevant to both hardware and software.

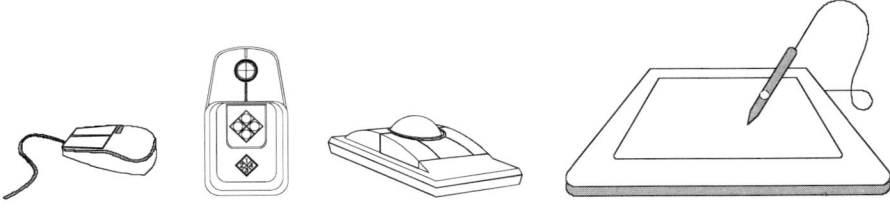

Part 9: *Requirements for non-keyboard input devices* (CD) specifies requirements for the design and usability of input devices other than keyboards, such as:

- mice
- pucks
- joysticks
- trackballs
- tablet overlays
- touch-sensitive screens
- light-pen and styli
- thumb-wheels
- hand-held scanners
- hand-held bar code readers
- remote-control mice

The aim of this part is to influence the design of these devices so that they accommodate the user's biomechanic capabilities and limitations and allow adequate safety and comfort. Part 9 specifies requirements for input devices commonly used for tasks such as:

- pointing
- selecting
- dragging
- scanning
- free-hand input

Part 9 provides guidance on the physical characteristics of input devices, including the force required to operate them, as well as their feedback, shape, and labeling.

General guiding principles are included in Part 9 and are the basis upon which the design requirements are based. This part also includes general requirements that apply to all input devices as well as specific requirements for individual devices. However, it does not include specific requirements for scanners, bar code readers, and remotely controlled mice. The requirements in this standard part do not cover the following input devices:

- eye-trackers
- head-mounted controllers
- speech activators
- input devices for disabled users

A unique feature of Part 9 is the requirement that a user must be able to use the device with his or her hands in a neutral position. This requirement will have a major impact on traditionally designed input devices that are operated with the hand deviated from a neutral posture.

Part 10: *Dialogue principles* (DIS) specifies a set of high-level dialogue design principles for command languages, direct manipulation, and form-based entries. It is intended to promote a user-centered approach to the development and evaluation of dialogue systems.

The objective of Part 10 is to optimize dialogue design in terms of its effectiveness, efficiency, and satisfaction. Its specifications are intended to provide assistance to users with different levels of competence, knowledge of work procedures, familiarity with a system, and frequency of use. It also requires consideration of user's different cognitive abilities in terms of attention span, short-term memory, learning, capacity, and interaction with a dialogue system. It requires that users be informed of the task requirements and qualifications necessary for interface with the dialogue.

Part 11: *Guidance on usability* (DIS) explains the way in which the user, equipment, task, and environment should be described—as part of the total system— and how usability can be specified and evaluated. It addresses usability in terms of meeting users' needs, provides a basis for measuring and specifying usability for ergonomic comparisons of products, and aids in identifying the context of use that is relevant for determining the applicability of the recommendations in other parts of 9241.

Part 11 states that usability is to be described and measured by the following characteristics:

- system *effectiveness*—the extent to which the intended goals of using the overall system can be achieved
- *efficiency*—the resources that have to be expended to achieve the intended usability goals
- *satisfaction*—the extent to which the user finds the overall system acceptable

The measures are required to be in a form that is sufficiently precise to allow them to be verified and the essential characteristics of the context to be reproduced. Specifications for usability measurement are to consist of target or actual values of these measures and a description of the relevant context of use.

This part states that to design and/or evaluate the attributes of hardware and software requires knowledge of the characteristics of the context in which the product will be used. It requires that a description of the relevant components of system be provided.

Part 12: *Presentation of information* (CD) specifies requirements for the coding and formatting of information on computer screens. It also gives guidance to designers of dialogue systems on how to apply usability principles to the arrangement and structure of information on the display.

The requirements of Part 12 are based on the results of physiological and psychological testing of user information processing. Part 12 states that in order to achieve optimum information processing, the information output is required to

- be properly adapted to the physiology of the human sensory organs (e.g., vision, hearing, touch, etc.)
- present the maximum amount of information appropriate for the task and the users' sensory and information-processing capabilities
- facilitate correct recognition and processing of information presented
- avoid unnecessary and excessive stress to the user

Coding is the assignment of an image to a specific meaning. To comply with this part, coding must be: clear, meaningful, conventional, easy to learn, easy to remember, easy to use, quick and accurate to use, non-distracting, and consistent.

Formatting concerns the arrangement and structure of the information on the screen. This part states that formatting should ensure that the interface matches the task requirements, optimizes users' performance (in terms of speed and errors of information recognition, reading and interpretation) and minimizes stress.

Part 13: *User guidance* (CD) specifies requirements and attributes to be considered in the design and evaluation of the software user interfaces. Its main purpose is to:

- aid the interaction with the system
- promote efficient system use
- avoid unnecessary mental workload
- provide support in managing error situations
- provide support for different skill levels

The recommendations in this part correspond to common situations (such as loss of control and need for help) involving special needs for information and actions.

This part addresses explicit user guidance and includes information explaining current objects or commands, additional information available on user request (such as help texts), and system messages. On-line tutorials, manuals, and user guidance outside of the system are not included.

Part 14: *Menu dialogues* (DIS) provides conditional requirements and recommendations for menus in user-computer dialogues. The conditional requirements and recommendations cover menus presented by various features including windowing, panels, buttons, and fields. The menu requirements relate to either: dialogue design, input design, or output design.

Dialogue design determines the way in which a user is guided by the system and influences the amount of control the user has over the dialogue. Part 14 requires that the dialogue be designed to support users during their work without being hindered by the additional work caused by system peculiarities. It states that the dialogue should enable the user to become well-informed and stay in control of the flow of work. These design goals are to be considered in the design of the menu structure, menu navigation, and methods for selecting menu options.

Requirements for input and output design are based on perception, identification, and discrimination of information. The input rules in this part describe how input devices can be applied to facilitate the input of data into the system. The devices provided are to depend on the task requirements and user's preferences. The output rules describe how data should be presented consistently and distinctly on the screen. These rules describe the placement of menu options and option groups, the structure and syntax for textual, graphic, and voice options, and presentation techniques to indicate option accessibility and discrimination.

Part 15: *Command dialogues* (CD) provide conditional recommendations for command languages. Like Part 14, this standard part requires that command dialogues be designed to support the user's work without being hindered by additional work caused by system peculiarities. It also specifies that the command dialogues enable the user to become well-informed and to allow the user to maintain control of the work flow.

The structure and syntax requirements of this standard part include specifications to maintain internal consistency for command macros, argument and syntax structures, command separation, language correspondence, command arguments, and quantifiers. Command representation specifications include command names, abbreviations, and function and hot keys. Input and output considerations include command reuse, command queuing, error correction, editing, misspellings, defaults, destructive commands, customization, echoing and output control, and format. Feedback and help requirements include command acceptance, error feedback, error highlighting, command information, performance aids, and parameter lists.

Part 16: *Direct manipulation dialogues* (CD) provides guidance on the design of manipulation dialogues in which the user directly acts upon objects or object representations (icons) to be manipulated. Direct manipulation dialogues are often used simultaneously with graphical user interfaces (GUIs). The manipulations are controlled by an input device and include:

- ◆ pointing at objects
- ◆ moving objects
- ◆ changing the physical characteristics of images

The manipulated objects that are included in the specifications in Part 16 are:

- concrete
- graphic
- representative of abstract software structures or capabilities

This part describes typical applications, selection criteria, and design requirements for graphical user interfaces. It includes display technology considerations, input device considerations, syntax of user operations, visual representation of objects and manipulations, user orientation and memory aids, feedback and help, and expert user considerations.

Part 17: *Form filling dialogues* (CD) consists of a number of conditional recommendations concerning the structure and semantics of dialogues in which the user enters or modifies fields in a specific field presented by the system. The recommendations of this part are to be met within the specific context for which they are relevant (that is, for particular users, tasks, environments, and technologies).

This part includes the use of non-text methods for providing form entries (that is, list boxes) and pertains to dialogue boxes that utilize form-filling dialogue techniques. Applications of this part include processing of insurance forms, income tax forms, medical forms, and purchase orders. Specifications in the part include:

- screen density
- field layout and size
- cursor movement
- list boxes
- scrolling lists
- choice buttons
- navigation

Application Domains

ISO 9241 was originally intended to apply to text and simple graphic office tasks. In 1990, ISO TC 159/SC4 extended the domain of 9241 beyond office settings to include other work environments in which VDTs are used for the types of tasks for which the original ISO standard was developed. These extended domains include public access, telecommunication, engineering applications, control rooms, medical and scientific applications, and office work. Some of the ISO 9241 parts apply to these domains, others need either substantial or minor changes to apply; a few of the standard parts do not apply to these domains (see Table 3–4). For example, for public access, most of 9241 applies, but there are other important issues—such as specifications for products infrequently used or used by semi-skilled users—that are not currently under its scope. In addition, user interface for the education domain is an important area that should be covered by an ergonomic standard, but there has been considerable debate by

Table 3–4 Application domains of ISO 9241

Domain	Part													
	1	2	3	4	5	6	7	8	9	10	11	12	13	14-17
Office work	✓	✓	✓	✓	✓ δ	✓	✓	✓	✓	✓	✓	✓	✓	✓
Computer aided design	Δ	δ	Δ	Δ	δ	Δ	✓	Δ	✓	✓	δ	Δ	Δ	δ
Advanced manufacturing technology	Δ	δ	Δ	Δ	δ	Δ	✓	Δ	✓	✓	δ	✓	Δ	δ
Medical	δ	✓	Δ	Δ	Δ	Δ	δ	Δ	✓	✓	δ	Δ	Δ	δ
Telecommunication	δ	✓	✓	Δ	Δ	Δ	δ	✓	✓	✓	δ	✓	Δ	δ
Control rooms	∅	Δ	δ	Δ	Δ	Δ	δ	Δ	✓	✓	δ	Δ	Δ	δ
Public access	δ	∅	Δ	∅	∅	∅	✓	✓	✓	✓	δ	✓	✓	δ

Legend:
✓ = already applies
Δ = substantial changes required
δ = minor changes required
∅ = does not apply

the ISO Displays and Controls technical committee about the circumstances under which hardware parts should apply for this area.

Status

Because each part of 9241 was started at a different time, each one is at a different stage of development. Some have been voted International Standards (ISs), some are still Committee Drafts (CDs), and some are Draft International Standards (DISs) (see Table 3–5).

Content Expansion

Although the intent of the original ISO 9241 standard was to make it independent of changes in technology, Part 3 focused on light emissive CRT displays, the most commonly used technology

Table 3–5 ISO 9241 parts and status

Part No.	Subject	Status
1	General information	IS
2	Task requirements	IS
3	Displays	IS
4	Keyboards	DIS
5	Workstation and posture	DIS
6	Environment	CD
7	Reflections and glare	CD
8	Color	DIS
9	Non-keyboard input devices	CD
10	Dialogue principles	DIS
11	Usability guidance	DIS
12	Presentation of information	CD
13	User guidance	CD
14	Menu dialogues	DIS
15	Command dialogues	CD
16	Direct manipulation dialogues	CD
17	Form-filling dialogues	CD

Table 3-6 New additions to ergonomic computer standards

Standard No.	Subject	Status
13406-2	Flat panel displays	CD
[not assigned]	Portable computers	WG

for office displays at that time. Because of the increase in use of other display technologies, ISO TC 159/SC4 decided to create a new ergonomic display standard (ISO 13406-2) that would accommodate displays with like liquid crystal and electroluminance technologies. Because of the increase in these technologies for lap top computers, ISO is also considering creating a separate ergonomic standard for portable computers (see Table 3–6).

Other Applications

Within each part of 9241, there are many specifications that can apply to areas and features beyond the scope of the standard. This standard is thus useful for designers and evaluators of prod-

Table 3-7 Part 3 requirements useful in control panel design

Product Feature	Control Panel Application
Viewing distance	✓
Viewing angle	✓
Character height	✓
Stroke width	✓
Width-height ratio	✓
Luminance modulation	N.A.
Character format	✓
Size uniformity	✓
Space between characters	✓
Space between words	✓
Space between lines	✓
Linearity	N.A.
Luminance	✓
Contrast	✓
Luminance coding	✓

N.A. = Not applicable

ucts other than VDTs and their peripherals. For example, although the characteristics of labels and controls on the control panels of printers, plotters, fax machines, and typewriter displays are not specified in 9241, many of the requirements of Part 3 can be used for them (see Table 3–7).

Examples of these requirements and recommendations are described in Table 3–8.

Table 3–8 Examples of requirements in ISO 9241

Part	Topic	Requirement
3	Character shape	A 7 × 9 (width to height) character matrix shall be the minimum used for tasks that require continuous reading. For non-dot-matrix techniques, equivalent character shapes should be achieved.
4	Auditory signals	The auditory signal should be perceptible in a normal office environment. The auditory signal shall be in the form of a click.
5	Seat height	Seat height shall be sufficient to prevent spinal flexion. The over-riding consideration is that seat height should be appropriate to that of associated working surfaces.
6	Thermal conditions	Local heat build-up due to thermal radiation or hot air should be avoided by suitable control of the climatic conditions. The working persons shall be protected from thermal stress from equipment.
7	Luminance ratio	From the standpoint of legibility, the luminance ratio of the image, including superimposed specular and diffused reflected luminance, shall be equal to or greater than 3:1. From the point of user acceptance, the luminance ratio of a specular reflection against the screen background should be low, compared to the image luminance ratio without specular reflections.
8	Color character size	Where accurate color identification of alphanumeric character strings and data entry fields are required, the character height shall subtend at least 20 minutes of arc at the design viewing distance. Where accurate color

Table 3–8 *Continued*

Part	Topic	Requirement
		identification of an isolated image (for example, a character or a symbol) is required, the image should subtend at least 30 minutes of arc at the design viewing distance, and preferably, 45 minutes of arc.
9	Lightpen size	Cylindrical lightpens shall be between 120 mm to 180 mm in length and 7 mm to 20 mm in diameter. Any other shapes of stylus or lightpens should be of sufficient size to ensure acceptable performance and minimize pressure points and biomechanical load.
10	Dialogue undo	If interactions are reversible and the task permits, it should be possible to undo the last dialogue step (from Part 10).
12	Color coding	Each color code should only represent one category of displayed data. Symbols shall be used consistently for the same meaning or function.
13	System response	Every input by the user should produce perceptible feedback from the system. System response to user entries should be fast enough so that it does not distract the user from the task.
14	Menu titles	If menu panels contain multiple menus or option groups, each menu option group should be titled . . . and these titles shall be perceptually distinct from the option.
15	Error feedback	Error feedback should be provided after the full command has been entered rather than as soon as the error is discovered by the computer.
16	Window status	If user input is not possible at a given time, visual cues shall indicate this system state. When a window is restored by the user, it should appear in the same location on the display and have the same size that it had when it was closed.
17	Windows, cursors & pointers	A unique window identification area should be provided for each primary window. Cursors and pointers used for different functions should be visually distinctive.

Compliance

Mandatory requirements of ISO 9241 are indicated by inclusion of a "shall" in the specification statement(s) for each requirement; recommended requirements are indicated by a "should" in the specification statement(s). All the hardware parts contain both mandatory requirements and recommendations; most of the software parts contain only recommendations (see Table 3–9).

Compliance Assessment

The seventeen parts of ISO 9241 contain not only specifications but also measurement metrics. The required measurements for parts that are hardware oriented are objective and quantifiable. Some of the hardware parts contain alternative test methods for products that do not meet the standards minimum specifications. These alternative test methods include evaluation procedures for

Table 3–9 Compliance specifications in Parts 1–17

Part No.	Subject	Requirement	Specification
		Should	Shall
1	General information		
2	Task requirements	✓	
3	Displays	✓	✓
4	Keyboards	✓	✓
5	Workstation and posture	✓	✓
6	Environment	✓	✓
7	Reflections and glare	✓	✓
8	Color	✓	✓
9	Non-keyboard input devices	✓	✓
10	Dialogue principles	✓	✓
11	Usability guidance		
12	Presentation of information	✓	
13	User guidance	✓	
14	Menu dialogues	✓	✓
15	Command dialogues	✓	
16	Direct manipulation dialogues	✓	✓
17	Form filling dialogues	✓	

user performance and comfort. Measurements for the parts that are primarily software can be measured by observational as well as empirical analysis (see Chapter 9).

User performance measures in ISO 9241 include speed and accuracy; comfort and ease-of-use assessments are determined by the user ratings (see Table 3–10). Several parts (e.g., Part 3 (Displays), Part 4 (Keyboards), and Part 9 (Non-keyboard Input Devices)) have sections that specifically describe user performance measurement procedures. Part 4 contains a general statement regarding the measurement of biomechanical stress; however, Part 9 specifically describes biomechanical measurements.

Table 3–10 Compliance assessments in ISO 9241

Parts	Direct Measures	Observations	Alternative Compliance Tests		
			User Performance	Comfort Assessment	Biomechanical Assessment
1					
2					
3	✓			✓	
4			✓	✓	✓
5	✓			✓	
6	✓			✓	
7	✓			✓	
8	✓		✓	✓	
9	✓		✓	✓	✓
10		✓			
11		✓			
12		✓			
13		✓			
14		✓			
15		✓			
16		✓			
17		✓			

Demonstrating Compliance

In order to claim product or environmental compliance with ISO 9241, evidence must be obtained demonstrating that its requirements are met using the measurement techniques specified in each part. Some of the protocols measure physical features like character size or key displacement; others determine the existence of a feature such as a detached keyboard or a display contrast and brightness control (see Table 3–11). Thus some of the compliance measurement procedures require sophisticated test instruments such as micro-photometers and luminance meters; others require visual verification through observations or empirical analysis (see Chapter 9).

Table 3–11 Example of compliance methods

		Measurement Method	
Part	Subject	Observation	Instrumentation
1	General information		
2	Task requirements		
3	Displays		✓
4	Keyboards		✓
5	Workstation and posture	✓	
6	Environments		✓
7	Reflections and glare		✓
8	Color		✓
9	Non-keyboard input devices	✓	
10	Dialogue principles	✓	
11	Usability guidance	✓	
12	Presentation of information	✓	
13	User guidance	✓	
14	Menu dialogues	✓	
15	Command dialogues	✓	
16	Direct manipulation dialogues	✓	
17	Form filling dialogues	✓	

Alternative Compliance Methods

An alternative way to demonstrate compliance is available for products that do not comply with the design specifications of parts of 9241. If it can be demonstrated through controlled usability testing that user performance, comfort, and preferences for a non-complying product are at least as good as a product that meets the standard, compliance can be claimed. If alternative compliance testing does not meet this criteria, then product manufacturers can redesign the product to the specifications of the standard or not market the product in the regions (country or market) requiring compliance with the standard.

Summary

ISO 9241 specifies requirements for all the components of a computer display workstation. The standard contains both normative and non-normative requirements. It includes specifications for hardware and software components of VDTs as well as requirements for related furniture and equipment and environmental conditions. This standard also contains a number of compliance test methods (see Chapter 9).

Reference

1. The text is a condensation of descriptions in ISO 9241, Parts 1–17.

Chapter 4

European Ergonomic Requirements

Overview

This chapter provides an overview of the ergonomic standards and legislation in Europe. It begins with a description of the two major economic European organizations—the European Union (EU), formerly known as the European Economic Community—and the European Free Trade Association (EFTA). An explanation of the EU's standards process is followed by descriptions of the ergonomic directives of the EU and the European standards organization (CEN). A summary of ergonomic standards activities of several European countries is then provided to demonstrate their similarities and differences.

The EU and EFTA

European standards activities are important to the global market because Europe contains a large population, capital, and GNP (gross national product). It is now the largest free trade zone in

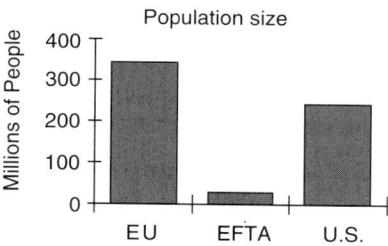 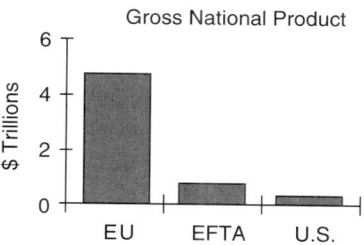

Figure 4–1 GNP and population of EU and EFTA compared to the United States

the world. The EU member countries contain more people (341.9 million), more capital and a greater GNP ($ U.S. 4.7 trillion) than North America. EFTA consists of an additional 32.5 million people and $U.S. 800 billion GNP (see Figure 4–1).

The European Economic Community (EEC) was established by the Treaty of Rome in 1957 and became operative in 1958. The general goal of the EEC was to create a common market for its member nations, specifically to remove physical, fiscal, and technical barriers between its member nations. Examples of physical barriers are border and custom controls; fiscal barriers include excise and value-added taxes; technical barriers are caused by exclusions of specific technologies, such as cathode ray tube displays or flat panel displays. The EEC is now called the European Union (EU) and is composed of fifteen members (see Figure 4–2): Belgium, Denmark, Germany, France, Greece, Ireland, Italy, Luxembourg, Netherlands, Portugal, Spain, the United Kingdom, Sweden, Finland, and Austria. Turkey has applied for admission.

The European Free Trade Association (EFTA) is similar to the EU in that it is an economic consortium, but it is now only composed of four countries (see Figure 4–2): Iceland, Norway, Switzerland, and Liechtenstein. EFTA was established in 1959 and became operative in 1960. EFTA was organized to remove trade barriers between its members countries. However, each member country was to maintain its own commercial policy toward countries outside the group. In 1977, EFTA entered into an agreement with the EEC to establish industrial free trade between member countries of the two organizations.

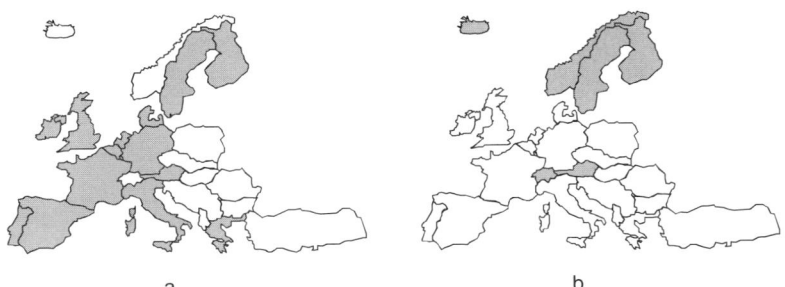

Figure 4-2 Members of the EU (a) and EFTA (b) (shown by shaded areas)

EEA

The European Economic Area (EEA) is a single market, joint trading block between EFTA and the EU established by the 1992 *Agreement on the European Economic Area.* It essentially provides for the extension of the EU's single market to the EFTA states that undertake to incorporate into their national laws EU's accumulated legislation on the free movement of goods, services, people, and capital. The agreement also provides for cooperation on social policy, consumer protection, environmental matters, research, and education.

National Standards Agencies

All European countries have national standards agencies. However, some participate ("P" members) and vote on standards, whereas others, known as observer members ("O" members), only receive information on standards. The kind of membership assignment depends on the choice of each country. For example, only 14 countries vote on ISO 9241 (see Table 4–1).

European Standards

The two European standards organizations are the Center for European Standardization (CEN) and the European Committee for Electrotechnical Standardization (CENELEC). The CEN and CENELEC standards development processes are similar to that

European Standards

Table 4–1 European standard organizations voting on ISO 9241

Country	Standards Agency	Participant Voting Members	Observer Non-voting Members
Albania	KCSA		
Austria	ON	3	
Belgium	IBN	3	
Bulgaria	BDS	3	
Cyprus	CYS		3
Czech Republic	CSN	3	
Denmark	DS		3
Finland	SFS	3	
France	AFNOR	3	
Germany	DIN	3	
Greece	ELOT	3	
Hungary	MSZH	3	
Ireland	NSAI		3
Italy	UNI	3	3
Netherlands	NNI	3	
Norway	NSF	3	
Poland	PKNMJ	3	
Portugal	IPO		3
Romania	IRS		3
Spain	AENOR	3	
Sweden	SSD	3	
Switzerland	SNV		3
Turkey	TSE		3
United Kingdom	BSI	3	
Yugoslavia	SZS		3

of ISO (see Chapter 3). Their standards are created through a consensus process among the EU member countries. Once the content for a new standard has been developed by an officially sanctioned technical committee (TC), it is circulated to the countries for comments. The TC reviews all comments, makes appropriate changes, and circulates draft standards among national members for a weighted majority vote. European countries are committed to adopt as their national standards the common stan-

Figure 4–3 Rendition of CE approval mark

dards developed by CEN and CENELEC and to withdraw any conflicting national standards.

Products that meet CEN standards are given a CE symbol to indicate they are in compliance with these standards (see Figure 4–3). The symbol is placed on the back or bottom of the product. Products that conform to the CEN standard and display the CE mark have the right to be put on sale anywhere in Europe. Those products that have been given the CE mark, but are later deemed to be unsafe are required to be removed from the European market and publish the reasons for their removal. The European Commission will then communicate this information to concerned parties and may inform all the European members of the action. In some cases, the action may be brought before the European Court of Justice.

In order to demonstrate compliance with a CEN standard, a product must present one of the following:

- a report describing compliance testing results by an independent body
- a compliance certificate issued by an independent testing agency
- a manufacturer's declaration of test results (supported by data from an independent agency or the manufacturer's test results)

Thus far, CEN has not developed specifics on testing and certification procedures for meeting CEN requirements. In addition, Europe has not yet decided whether non-European testing and certification laboratories can receive full European accreditation or whether their test results will be accepted as valid.

CEN has committed to adopt ISO 9241 as its standard. National standards will subsequently be superseded (that is, replaced) by ISO 9241.

European Directives

When requirements are legislated by the EU, they may become directives. A directive is one of three types of binding legal instruments for implementing policy or decisions (the other two are "regulation" and "decision"). A directive is an instruction from the European Council to one or more EU member states to legislate on a specific matter within a defined period of time. It provides an outline of the required legislation but leaves the details of implementation to the member states.

In the area of the EU directive for display screens, member states can use either their national standard or ISO standard for the implementation of the directive.

 **EU Directives**

The Social Charter

In 1989, the EU adopted a social charter which aimed to provide a common standard across its member countries for social rights and for employees' living and working conditions. The charter was intended to enable workers to obtain greater geographic and employment mobility. It addressed six major issues:

1. freedom of movement and employment throughout the EU
2. improved living and working conditions
3. right to free association and bargaining
4. health and safety in the workplace
5. right to vocational training
6. protection of special interest groups

The Safety and Health Directive

To enhance the EU's ability to improve working conditions and provide a healthy and safe workplace, the European Council of Ministers integrated the directive Health and Safety (Directive 391) into their social charter in 1990. This directive includes a Framework Directive (Article 16) that assigns general responsibilities to employers and employees. In addition, it contains a number of directives that relate to specific issues, such as:

1. workplaces
2. work equipment
3. personal protective equipment
4. manual handling of heavy loads
5. work with visual display equipment
6. work with carcinogens
7. work with biological agents

Display Screen Directive

The fifth directive—*Minimum Safety and Health Requirements for Work with Display Screen Equipment* (90/270/EEC)—was issued by the European Commission (an institution of the EU) in 1990. It specified minimum requirements, but not design limits, for visual display terminal hardware, software design, use, and the environment. It went into effect on January 1, 1993.

The goal of this directive was to ensure a minimum level of health and safety for VDT workers in EU member countries. It describes minimum health and safety requirements for VDTs, general guidelines on responsibilities, and also identifies areas for legislation. The directive applies to all office workstations used as a major part of the worker's task. It specifically addresses these topics:

- displays screens
- keyboards
- work surfaces
- work chairs
- environmental conditions
- [software] user interfaces

EU Directives

The directive requires employers to provide:

- information disclosing health implications of products and environments
- worker training in product use
- work breaks
- vision testing
- special viewing glasses
- appropriately designed equipment, furniture, work environment, and software

The requirements in this directive that relate to hardware design are listed in Table 4–2; those that relate to the work environment are in Table 4–3; requirements for software are in Table 4–4.

Table 4–2 Hardware requirements

Topic	Requirement
Displays	Clear characters
	Stable, flicker-free images
	Adjustable brightness and contrast
	Easy swivel and tilt
	Glare and reflection free
Keyboards	Tiltable and separate
	Arm/hand rest space
	Non-reflective, matte finish
	Adequate symbol contrast
Desks	Low surface reflectance
	Document holder
	Sufficient space to ensure comfort
Chairs	Stable
	Freedom of movement
	Seat height adjustable
	Back height and tilt adjustable
	Available footrest

Table 4–3 Environment requirements

Topic	Requirement
Space	Dimensioned and designed to allow change of posture
Lighting	Satisfactory lighting conditions
	Secondary adjustable lighting as appropriate
	Glare avoidance by layout of workplace/design of lighting
	Daylight control by suitable window covering
Reflection and glare	Protection from direct glare and minimization of screen reflections
Noise	Avoidance of distracting noises
Heat	No discomfort from equipment heat
Radiation	All electromagnetic radiation (except visible light) at negligible levels
Humidity	Adequate maintenance of levels

Table 4–4 Software requirements

Topic	Requirement
Suitability	Software suitability for the task
Usability	Easy-to-use software and, where appropriate to the operator's level of knowledge or experience; no quantitative or qualitative checking facility use without the knowledge of the workers
Feedback	Provision of feedback to workers on their performance
Format and pacing	Display of system information in a format and at a pace that are adapted to operators
Ergonomic principles	Application of the principles of software ergonomics

Directive Applications

The display screen directive specifically describes applications to which its requirements apply and do not apply (see Table 4–5). Although the display screen directive excludes portable computer systems not in prolonged use, it states that "The use of such equipment must not be a source of risk to workers." The directive does not provide measurable ergonomic specifications.

Table 4–5 Applications for the display-screen directive

Applications
Displays
Input devices
Software
Accessories (e.g., document holders and palm rests)
Peripherals (e.g., printers, plotters, disk drives, telephones, modems, desks and chairs)
Environment

Exempt Applications
Vehicle control panels
Transport computer systems
Public use computer systems
Portable systems for infrequent use
Small, data/measurement displays (e.g., calculators, cash registers)
Traditional typewriters

Although the display-screen directive is a requirement for employers, manufacturers and designers of VDT products and associated environments need to ensure compliance because their customers are employers who must comply with the directive. Thus, manufacturers who sell or wish to sell VDT products in the EU should expect to receive requests for information regarding usability, health and safety implications, and test results of their products from their customers and potential customers. Employers must conduct analyses of all VDT workstations to determine health and safety risks. All workstations installed before December 31, 1992, are required to meet the directive requirements by 1996. Workstations purchased after December 31, 1992, are to meet the minimum requirements in the directive.

National Legislation

The display-screen directive requires EU member countries to create laws and regulations to demonstrate compliance with the directive. This act of creating national ergonomic legislation is one of the important changes in the last decade regarding ergonomic requirements. It is the first time that a large, multinational market will have ergonomic laws in addition to standards for the design and use of computer products. The product design marketing and compliance implications for industry are extensive.

Most of the EU and EFTA member countries have now transposed the directive into national legislation. The Commission has commenced checking the conformity of their transpositions. In addition, the Commission has opened infringement procedures against those countries that have not accomplished the transposition.

CEN

CEN is the standard organization that issues common standards for the eighteen nations in the European Union and European Economic Area. The primary objectives of CEN are to promote

- openness and transparency
- consensus
- national commitment
- technical coherence at the European and national levels

CEN is responsible for the planning, drafting, and adoption of all standards except those pertaining to sectors of electrotechnology and telecommunications, which are the responsibility of CENELEC. CEN and CENELEC have agreed to adopt standards developed by other European standards organizations. When they are adopted, they become European norms (ENs) or standards for all member states. European norms define the technical standards that manufacturers need to meet in order to design and manufacture products for the European market. Under an agreement signed by the EU member states—the Treaty of Rome—all member states must adopt EN standards.

CEN creates a standard only when the need for a European standard has been established and when an existing standard cannot be used. A CEN standard is official only after it has been adopted in each of the eighteen CEN member countries. Compliance with CEN standards is voluntary. However, when a CEN standard is related to a European directive it may constitute a presumption of conformity to the directive.

Members of CEN are also members of ISO. The two organizations have a formal agreement—the Vienna Agreement—for liaison, cooperation, and information exchange. CEN is committed to use international standards whenever possible as a basis for its work. CEN creates standards only when European harmonization or standards development deadlines cannot be met.

Ergonomic standards relating to machinery are created by CEN/TC 122. The TC 122 WG 5 establishes ergonomic requirements for VDU workplaces. ISO and CEN are coordinating their efforts to create a VDT standard for Europe that harmonizes with ISO 9241. When CEN adopts an ISO part, it uses the number of the ISO standard with the prefix "EN". Thus ISO 9241 becomes EN 29241.

Germany

There are several organizations in Germany that create ergonomic regulations and standards (see Table 4–6). The regulations created by these agencies are enforced by the Federal Minister of Labor and Social Affairs. DIN safety standards are harmonized with international standards such as ISO and IEC.

In Germany, the Equipment Safety Law allows manufacturers or importers of technical equipment to display or distribute equipment only if users are protected against hazards to their

Table 4–6 German standard-development agencies

Agency	Abbreviation
German Institute for Standardization	DIN
German Electrotechnical Committee	VDE
Association of German Engineers	VDI

health or life during proper use. "Technical equipment" means working installations and equipment that can be properly used without the addition of further parts or accessories. "Proper use" means the use of equipment as specified by the manufacturer or normal use apparent by the construction and design of the equipment. Failure to observe safety regulations can result in legal action. In addition, manufacturers and importers may be prevented from selling and exhibiting unprotected technical equipment. However, according to a general clause in all German safety regulations, the regulations need not be applied if equal safety can be assured through other measures. The German safety law requires that operating instructions be bundled with the delivery of equipment.

Besides the equipment safety law, a number of other laws and regulations exist that apply to safety and use of visual display terminals. Examples include

- a regulation of electrical equipment use in an explosive atmosphere
- an x-ray act that applies to all equipment in which electrons are accelerated in excess of 5 kV
- an emission protection law that restricts use of polyvinyl chloride in capacitors and transformers

Germany's first ergonomic ordinance was Safety Rules for Office Workstations (ZH 1/618), which specified regulations that applied to VDT operations. In 1980, the Safety Regulations for Display Workplaces in the Office Sector (ZH 1/618) was published. ZH 1/618 is an ordinance, not a federal law or regulation. However, it is legally binding because failure to observe safety regulations can result in legal action.

More than one German regulation or standard can apply to a product. For example, a visual display terminal must comply with the following:

- electrical standard—DIN IEC 950/EN 60950
- implosion standard—DIN IEC 65/VDE 0806
- ergonomic ordinance—ZH 1/618

Legal Aspects

Although all products sold in the German market have to comply with the Equipment Safety Law, distributors or manufacturers are not fined if their products do not comply. Instead, German authorities publish an order forbidding the distribution of noncomplying products. In addition, inspectors for worker compensation insurance can demand modification of equipment and forbid its use until correction.

All German employers must insure their employees in a worker compensation association. Worker compensation insurance associations (Berufagenossen-schaften) are independent agencies. The associations pay for treatment of work injuries and for pensions in case of death or occupational disease. The associations are required to create rules to reduce work accidents and conduct inspections to assure that the rules are followed. Accident prevention rules are legally binding and result in fines when employers and employees do not follow them. Safety rules are recommended practices which do not necessarily result in fines if not followed.

Officers from industrial inspection boards and technical experts from the accident insurance institutions are considered to have the required technical knowledge to judge the safety of the technical equipment sold or exhibited by manufacturers. However, product testing is conducted by testing agencies (see Figure 4–4) which give certification labels to products that pass their tests. These agencies are approved by the Federal Ministry of Labor and are authorized to issue safety marks (see Figure 4–4).

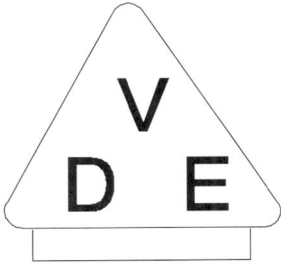

Figure 4–4   Rendition of test label for VDE standards agency

Demonstrating Compliance

There are more than sixty agencies authorized to test equipment. All test agencies are supposed to perform the same standardized tests. Manufacturers may choose an agency to test their product or they may choose to prove compliance themselves. Certain work areas are assigned to each test center, but various agencies can test the same product categories. Some agencies (such as VDE and TUV) have established test laboratories in other countries.

Tested products that comply with German standards are marked with a "GS" safety mark (see Figure 4–5a) which stands for "safety tested." The GS mark includes the label of the test agency (refer to Figure 4–4) in the upper left corner (see Figure 4–5b) or on the left side when the GS mark is less than 2 cm in height. Some test agencies only authorize the safety mark for three years and require retesting.

DIN 66234

The most influential German ergonomic standard has been DIN (Deutsches Institute fur Normung) 666234: *VDT Workstations*. DIN 666234 consists of the following 10 parts:

Part 1: Geometric design of characters includes requirements for characters having high edge definition such as those in vacuum fluorescent and liquid crystal displays. It applies to displays in VDU workstations used for reading or

Figure 4–5 Rendition of German safety (GS) test mark (a) and GS mark combined with VDE test agency (b)

combined reading and typing of medium and long duration in well lighted offices.

Part 2: Perceptibility of characters on screens specifies requirements for bright characters on dark background screens. It does not include liquid crystal displays and vacuum fluorescent displays.

Part 3: Grouping and formatting of data provides general application guidelines, explanations, and examples of grouping and formatting of the following:
- the arrangement of dedicated areas on the screen
- user guidance
- field structure, including formal aspects and contents

Part 4: (Does not exist as it was reserved for a standard that was not developed.)

Part 5: Coding of information describes alternative graph presentations and appropriate color combinations for characters and backgrounds.

Part 6: Workstation design specifies furniture and arrangement of displays, keyboards, and documents.

Part 7: Ergonomic design of environment lighting and arrangement specifies lighting for work environments in which both positive and negative polarity displays are viewed.

Part 8: Principles of man-machine dialogue design is a set of guidelines based on these five principles:
1. suitability to the task
2. self-explanatory
3. controllability
4. conformance to user expectations
5. error tolerance and transparency

These guiding principles are important because they have become integrated into almost all of the ISO 9241 software parts and some of the hardware parts (see Chapter 3). Part 8 specifies work procedures, including how user-system interfaces should be designed from the point of view of the users' command language and related topics such as training. The primary aim of this part is to ensure that the dialogue is adapted to the psychological processes of users.

Unlike other parts of DIN 66234, Part 8 contains no conformance requirements.

Part 9: Measuring methods describes how the values specified in the other parts of 66234 are to be measured. It includes measurement methods for laboratory and field testing during product development and after product completion.

Part 10: Minimum information to be specified was created to assist buyers in the comparison of products. It is in the form of a questionnaire which must be completed by the seller of the product. It specifies that information must be provided on these VDT characteristics:

- reflectance and glare
- filters
- screen radius
- height and tilt adjustability
- screen dimensions
- resolution
- number of character lines
- number of lines
- font size
- character height, width,
- spacing, generation type,
- color, luminance
- background color and luminance
- refresh rate

The publication of these ten parts (see Table 4–7) began in 1980. Almost all parts were published prior to 1990.

Table 4–7 Publication schedule of DIN 66234

Part	'80	'81	'82	'83	'84	'85	'86	'87	'88	'89	'90	'91	'92	'93	'94
1	✓														
2				✓											
3												✓			
5		✓													
6					✓										
7					✓										
8					✓										
9							✓								
10															✓

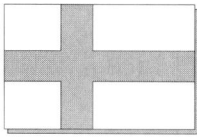

Sweden

Like Germany, Sweden played a major role in influencing international ergonomic standard development and European product marketing. With the creation of the Work Environment Ordinance of 1977, Sweden was the first country to develop requirements for work with display screens in the office work environment. This regulation established requirements for:

- displays
- working height
- vision testing
- employee transfers
- avoidance of monotonous work

In 1993, an ordinance was created to translate and implement the EU display screen directive. The ordinance specifies:

- the design of the VDU and keyboard
- lighting conditions and design of the workplace
- health and comfort protection from noise, heat, electric, and magnetic emission
- software and systems relating to user tasks, capabilities, and demands
- worker knowledge of quantitative and qualitative monitoring

The MPR

The Swedish National Board for Measurement and Testing (MPR) establishes product testing methods and accredits laboratories to conduct VDU testing when they fulfill MPR testing requirements. The accreditation is a duplication of the European standards requirements (EN 45001 and 45002), which also specify requirements to be met by testing laboratories seeking accreditation.

MPR 1. In 1985, the Swedish government ordered the Board for Measurement and Testing (MPR) to establish a system for testing VDUs. The purpose of this order was to provide user organizations (trade unions and health organizations) a tool for evaluating the health implications of VDUs. The order was mandated because of increased worker anxiety about the health effects of using VDTs.

The MPR published its first set of testing methods for VDTs—*MPR 1*—in 1987. The MPR development committee consisted of representatives from trade unions, employers organizations, computer distributors and manufacturers, health organizations, research institutes, and test laboratories. Two groups were formed from this committee: one to develop test methods for radiation emissions and the other to develop methods to evaluate ergonomic characteristics of VDTs (see Table 4–8).

Because of the concern about the possible health hazards of electromagnetic emissions from electronic equipment (especially VDUs and related equipment), the MPR 1 specified radiation emission limits and test methods even though there was insufficient medical and scientific evidence to use as a reference for emission limits. The limits were set as a precautionary measure and stipulated VDUs should not appreciably add to pre-existing electromagnetic ambient levels.

MPR 2. MPR decided that this initial set of test requirements was too diversified and expensive and the test results too difficult to interpret. After three years of use, MPR revised the requirements and published *MPR 2* in 1990. *MPR 2* (SWEDAC) consisted of a reduced set of product specifications, new emission characteristics, and revised test methods based on new VDU technologies. The test methods pertaining to basic principles of display image perception were retained and the test methods to evaluate display image quality were simplified. The new test methods are now included in two documents: *Test Methods for Visual Display Units* (MPR 1990: 8) and *User's Handbook for Evaluating Visual Display Units* (MPR 1990:10). Conversion of *MPR 2* into a CEN standard is being considered by CENELEC.

Table 4–8 Display features specified and tested by MPR

Specific Display Characteristics	Specification	Specific Display Characteristics	Specification
Display feature:		*Emissions:*	
Screen/cursor luminance	≥ 100 cd/m^2	Electrostatic potential	± 500 V
Luminance uniformity	$\geq 80\%$	Magnetic field:	
Reflex sensitivity, specular	$\leq 1\%$	2 kHz–400 kHz	≤ 25 nT, 50 cm around VDU
Reflectance, diffuse	$\leq 10\%$	5 Hz–2 kHz	≤ 250 nT, 50 cm around VDU
Jitter	0.0002 mm/mm reading distance	*Alternating electric fields* V/m:	
Critical flicker (CFF)	95%	2 kHz–400 kHz	≤ 2 nT, 50 cm around VDU
Character sizes and distortion	$\leq 10\%$	5 Hz–2 kHz	≤ 25 nT, 50 cm from VDU
Linearity	$\leq 1\%$		
Orthogonality	$\leq 1\%$		
External and internal luminance modulation	$\geq 70\%$		
Angle-dependent luminance modulation	diffuse at 40° $\leq 25\%$		
Raster modulation	$\leq 15\%$		
Sharpness/MTF-analysis	$\geq 85\%$		

Sweden's Union Standard

The Confederation of Professional Employees (TCO) is the dominant labor organization in the Swedish service sector and management work force with a membership population of 1.3 million—25 percent of Sweden's total labor force. In 1986, the TCO published the *Screen Checker,* a guideline that includes requirements for displays (see Table 4–9) and keyboards (see Table 4–10) as well as limits for radiation emissions. The TCO represents 85

Table 4–9 TCO specifications for displays

Display Specification	Minimum	Preferred
Screen size	12″	14″
Flicker	none	
Polarity		Positive
Characters	sharp & clear	
Color		black on light screens/ yellow and green on dark screens
Characters	all easily distinguishable	
Character height	3.8 mm	3.8–4.5 mm
Line distance	allows easy distinguishability	
Ghosting	none	
Reflections	does not impede readability	
Screen framing	3 contrast points	
Screen tilt	5°	5°–20°
Vertical adjustment	520 mm	520 mm–370 mm

Table 4–10 TCO specifications for keyboards

Keyboard Specification	Minimum	Preferred
Keyboard	separate from display	
Hand support	front edge to allow proper support	
Stability	no sliding / wobbling	
Height	30 mm @ home row	
Angle	5° adjustable	5°–11°
Noise	not irritating	
Surface finish	matte and light surface	
Key size	12 mm	12 mm–15 mm
Interkey space	18 mm	12 mm–20 mm

percent of all white-collar and professional employees in Sweden. Its objective in publishing the *Screen Checker* was to provide specifications for features of displays and keyboards that the TCO felt were most important to user comfort, health and safety. The *Screen Checker* was to be used as a guide in evaluating the ergonomic quality of existing equipment and potential purchases, and during product development. The TCO has published a number of other "checking tools" such as *Screen Facts* and *Software Checker* to inform users about appropriate design features and to provide guidelines for evaluating and purchasing information technology products.

TCO '92

In 1992, TCO published *Environmental Labeling of Displays,* which requires that equipment meet strict specifications for low electric and magnetic fields while being energy efficient and meeting electrical and fire safety standards.

TCO '95

TCO expanded its demands in its 1995 specifications: *Requirements for Environmental Labeling of Personal Computers* and *Test Methods for Environmental Labeling of Person Computers.* The *Requirements* document is the most stringent set of requirements for personal computers yet developed. It specifies requirements for desktop and laptop personal computer displays, processors, keyboards, and future input devices. The *Test Methods* document defines requirements for verifying product characteristics. Verification of compliance of some product features must be conducted by a third party test agency authorized by the TCO. Verification of other product features are allowed by supplier test reports or certificates.

Certification Agencies and Symbols

The symbol for a product that is considered to be environmentally acceptable is a falcon, which is used on its 1995 certification label. The Swedish agency that tests and certifies electrical products is SEMKO. Its services include the certification of company quality assurance systems (like ISO 9000, see Chapter 10), environmental certification, and post-certification evaluations on TCO-labeled displays.

Denmark

Denmark has two ergonomic regulations that are a result of negotiations between employers and employees to implement the EU display screen directive. If a violation of the legislation occurs, the government will conduct an audit to evaluate the violation. If the auditor discovers the violation has occurred, a fine or jail sentence can be imposed. *Instructions about Working with VDUs* published by the Danish Work Environmental Service provides instructions on implementation of the EU directive. The Instructions include a number of specifications that are not included in the EU directive. For example, they specify that the hand and arm should be in a neutral position during use of VDTs, tables should be individually adjustable for seated and standing users, and employers are to provide employees with personal equipment (such as glasses for VDT viewing). The Instructions are to be applied for employees who work at VDTs more than two hours a day.

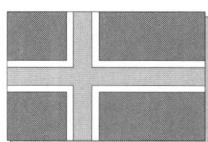

Norway

Norway is unique in Europe because it has voted not to become part of the EU. It will thus not be required to create a law to demonstrate compliance to the EU VDT directive.

However, Norway has an ergonomic regulation for VDT work—the Norwegian *Working Environment Act.* This regulation requires detachable keyboards, adjustable terminals, and a maximum of two-hour shifts of terminal work, and it limits the amount of time spent on terminal work up to 50 percent of the working day. It also provides for biannual eye tests and employer-provided glasses as well as on-the-job training in ergonomics.

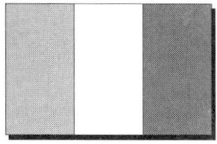 ## Italy

In 1968, Italy issued a law (#186/68) requiring that products must be designed to the state-of-art knowledge of function, health, and safety and also conform to the requirements of the Italian Electrotechnical Committee (CEI). Italy's first legislative decree (#626/94) for ergonomics was released in 1994 to interpret and enforce eight EU Directives including the Display-Screen Directive. The main requirements of the decree are:

- rules for employers, executives, companies, health and safety workers, representatives and workers
- equal treatment of health and safety and ergonomics
- assessment of the ergonomics and health and safety of all work places and work conditions
- formulation of assessments

Compliance to EN 29241 is necessary to meet 629/94. All new work places needed to comply by March 1995 and all existing work places need to comply by January 1, 1996. Violation of this law can result in a fine (up to $5,000 per violation) and a two- to six-month jail sentence.

 ## United Kingdom

U.K.'s VDT Regulations

The United Kingdom ergonomic regulations are described in the *Codes of Practice: Management of Health and Safety at Work* and *Display-Screen Equipment Work*. These codes were approved by the Health and Safety Commission under the 1974 *Health and Safety of Work Act*. These codes have implemented the requirements of the EU VDT directive and are thus directed at employers. They require eye exams, rest breaks, and training. There are seventeen regulations in the codes:

Regulation 1	Citation, commencement and interpretation
Regulation 2	Misapplication of these regulations
Regulation 3	Risk assessment
Regulation 4	Health and safety arrangements
Regulation 5	Health surveillance
Regulation 6	Health and safety assistance
Regulation 7	Procedures for serious and imminent danger
Regulation 8	Information for employees
Regulation 9	Co-operation and coordination
Regulation 10	Persons working in host employers' or self-employed persons' undertakings
Regulation 11	Capabilities and training
Regulation 12	Employees' duties
Regulation 13	Temporary workers
Regulation 14	Exemption certificates
Regulation 15	Exclusion of civil liability
Regulation 16	Extension outside Great Britain
Regulation 17	Modification of instrument

Requirements for display screen equipment work are described in the *Health and Safety (Display Screen Equipment) Regulations of 1992*. The guidance document for this regulation contains nine regulations and three annexes, including guidelines, a discussion of VDT health issues, and references:

Regulation 1	Citation, commencement, interpretation, and application
Regulation 2	Analysis of workstations
Regulation 3	Requirements for workstations
Regulation 4	Daily work routine of users
Regulation 5	Eyes and eyesight
Regulation 6	Provision of training
Regulation 7	Provision of information

Regulation 8   Exemption certificates
Regulation 9   Extension outside Great Britain
   Annex A   Guidelines on workstation minimum requirements
   Annex B   Display-screen equipment: possible effects on health
   Annex C   Sources of information

Annex A contains a description of the application of standards and a summary of U.K., ISO and CEN ergonomic standards. It specifies requirements for displays, keyboards, desks and work surfaces, chairs, lighting, glare, noise, heat, humidity, radiation, and software.

The contents of Annex B summarize research presented at ergonomic conferences in the 1980s and the report from WHO (see Chapter 2) and includes information on:

- upper-limb pains and discomfort
- eye and eyesight effects
- fatigue and stress
- epilepsy
- facial dermatitis
- electromagnetic radiation
- effects on pregnant women

United Kingdom VDT Standard

The 1990 British Standards Institution (BSI) standard on VDTs (BS 7179) contains six parts including requirements for display screens and keyboards used in office environments. This standard is intended to be an interim standard until CEN adopts ISO 9241. It was produced because the BSI recognized that industry needed guidance on ergonomic issues. It is intended for managers and supervisors as well as VDU equipment manufacturers.

ESPRIT

Since 1985, several of the research projects that have advanced the state-of-the-art of ergonomics and ergonomic standards have been funded by the European Strategic Program of Research and

Development (ESPRIT). ESPRIT is a ten-year transitional program to encourage pre-competitive research in Information Technology. Its programs bring together industrial and academic researchers in the EU to focus on research and development that will significantly contribute to the growth of the information technology industry in Europe. ESPRIT is funded jointly by European industry and the Commission of the European Communities (ECC).

The strategy of ESPRIT is to stimulate research in five major areas:

1. advanced microelectronics
2. software technology
3. computer integrated manufacturing
4. advanced information processing
5. office systems

Members of the European communities have combined university and corporate research and resources to integrate information systems through ESPRIT. This program provides funding for European research that enhances the quality of product user interface and thus enhances Europe's competitive edge as well. The goal of ESPRIT is to "focus European information technology industry upon the importance of usability and the human user for international success."

Summary

Although several European countries in the EU and EFTA have had VDT standards or ordinances for several years, these standards will be superseded as CEN adopts ISO 9241. ISO 9241 will then become the basic requirement for VDT furniture, hardware, software, and environments in Europe.

In addition, all the EU countries have, or will soon have, VDT legislation to comply with the EU display-screen directive. Some countries have created legislation with VDT and work practice specifications that are more strict than the EU directive.

Chapter 5

North American Ergonomic Standards

Overview

This chapter provides an overview of ergonomic standards and guidelines in the United States and Canada. These include both hardware and software requirements. Some of the basic differences in content and philosophy of industry software guidelines and standards of these two countries are described.

 ## United States Standards

The most widely referenced general ergonomic standards in the U.S. are *military standards* for equipment design (such as MIL-STD 1472) and the design process (such as MIL-H 46855) and *civilian* standards for the design of computer workstations (such as ANSI/HFS 100) and for facilities and environments as required for the disabled by the ADA (see Chapter 6).

U.S. Military Standards

The military establishment requires that systems, equipment, and facilities provide work environments that foster effective operations and personnel safety and health while optimizing performance and minimizing errors. In order to meet these goals and to design and obtain equipment that personnel can successfully operate, military requirements place stricter demands on designers and manufacturers than do other establishments. The military believes that its requirements minimize the time and cost of training and the risk of errors.

MIL STD 1472. *Human Engineering Design Criteria for Military Systems, Equipment and Facilities* (MIL STD 1472) specifies human engineering design criteria, principles, and practices that optimize the integration of military personnel into its systems, equipment, and facilities. The primary goal of 1472 is to optimize the success of military missions by creating systems that are simple, efficient, reliable, and safe.

MIL STD 1472 is based on research and operating experience in the military establishment. Thus, not all its requirements —particularly those related to anthropometry—apply to civilian population characteristics, products, and environments. In spite of this, the 1472 specifications have become a basis for many civilian standards including ANSI/HFS 100.

MIL STD 1472 includes requirements for

- visual displays
- audio displays
- control
- labeling
- anthropometry
- workspace design
- environment
- design for maintainer
- design of equipment for remote handling
- small systems and equipment
- operational and maintenance ground/shipboard vehicles
- hazards and safety
- aerospace vehicle compartments
- user-computer interface

The military requires that all of the systems they develop and the commercial equipment they purchase meet relevant specifications in 1472. If equipment to be purchased does not meet these specifications, it must be modified to comply.

MIL-H 46855. The *Human Engineering Requirements for Military Systems Equipment and Facilities* standard (MIL-H 46855) specifies design process responsibilities for procurement agencies and manufacturers. It includes analysis, documentation, design review, and testing required to ensure that a system will meet the needs of the procuring agency and the military establishment.

In addition to MIL STD 1472 and MIL-H 46855, each branch of the U.S. military service has ergonomic standards specific to its needs, equipment, and activities. Examples of standards from the various service branches are listed below:

- *Military Standardization Handbook: Human Factors Engineering Design for Army Material* (MIL-HDBK-759)
- *Air Force Human Factors Engineering Series* (AFSC DH 13)
- *Navy Defense System Software Development* (DOD-STD-2167) MIL-HDBK-759

MIL-HDBK-759. The *Military Standardization Handbook: Human Factors Engineering Design for Army Material* provides specifications and recommendations on a wider variety of topics than does MIL-STD 1472. The handbook includes specifications and measurements for a variety of applications, from environments with noxious substances and radiation to specific requirements for handles, fasteners, connectors, and packaging. It presents detailed specifications on controls, displays, consoles, workspaces, anthropometry, and body movement. It also includes several specifications on input devices, some of which were used in the Input Device section of the proposed revision of the 1988 ANSI/HFS 100 standard.

One section of this military handbook is the *Principles of Human Behavior Related to Safety*. It includes twenty-five principles about human behavior intended to help designers create safe equipment. Though basic and elementary, these principles (see Chapter 8, Section 1) are so important that they could be used as guiding principles and minimum requirements for all ergonomic standards, regulations, and guidelines.

U.S. Civilian Standards

ANSI/HFES 100. In 1983, American National Standards Institute (ANSI) approved a committee appointed and sponsored by the Human Factors Society to develop an ergonomic technical standard: *American National Standard for Human Factors Engineering of Visual Display Terminal Workstations*. This occurred in part as a result of the standards development in Germany (see Chapters 1 and 4) and the need of U.S. engineers, industrial hygienists, and personnel management for guidance on the design and selection of ergonomic features of VDTs and their workstations. The standard was also initiated to assure active participation of the U.S. in development efforts of ISO ergonomic standards.

Specifically, the ANSI VDT standard was expected to:

- codify areas of common agreement between scientists and human factors engineers
- provide guidance on the VDT workplace to designers and users
- address VDT equipment design and also common features of the VDT workplace and its environment
- represent the U.S. position in international ergonomic standard-setting activities for the next several years

The ANSI VDT standard committee consisted of seventeen members from the computer and furniture industry, academia, government, and consultant agencies. The members were required to be human factors scientists rather than individuals with a political, business, or social affiliation. The standard they created consisted of requirements for VDT environmental conditions, displays, keyboards, and furniture. The intent of the committee was to base the requirements on empirical evidence available from scientific research and principles agreed to by the human factors community.

Applications. The ANSI VDT standard was primarily developed for text processing, data entry, and data inquiry applications but was recommended for consideration for other types of workplaces and tasks in which visual display terminals or their components were used. It provided a technical base of information against which to evaluate computer workstations. It included a number of compliance measurements (see Tables 5–1 to 5–4) that manufacturers could use to test their products. Many of the requirements and measurements were then used as the basis for some of the requirements of Part 3 (*Display requirements*) of the ISO 9241 standard.

Alternative Compliance. The ANSI/HFS 100 standard was published in 1988. It offered an alternative way to demonstrate compliance if a product's design features did not meet the standard's requirements. The alternative method allowed a product to comply if it could be shown that the product could yield levels of human performance equivalent to or better than a system/product which met the standard, without a decrease in comfort. The general requirement to demonstrate compliance was that performance testing be conducted using standard research techniques under appropriate control conditions. However, the 1988 standard did not describe performance or comfort testing protocols.

Table 5–1 1988 HFS 100
Display requirement topics and measurements

Requirement Topic	Measurement
Luminance differences and uniformity	Display luminance, spot size, modulation, uniformity
Polarity	none
Color	Discrimination and legibility
Flicker and blinking	Flicker
Image alignment	Orthogonality
Image stability	Jitter
Glare guards	none
Characters	Size and uniformity
Viewing distance and angle	none
Controls	none

Table 5–2 1988 HFS 100
Keyboard features and measurements

Feature	Measurement
Layout	none
Cursors	none
Height and slope	Height and slope
Placement	none
Surfaces	none
Labels	none
Key shape and size	none
Key spacing	none
Stability	none
Key force and travel	Key force
Feedback	none
Profile	none

Table 5–3 1988 HFS 100
Furniture features and measurements

Feature	Measurement
Worksurface	
Clearance	Clearance under Work surface
Adjustments	none
Support surfaces	none
Chair	
Height	Seat reference point
Dimensions	Back height and seat depth
Angles	Seat angles
Backrests	none
Armrest	none
Chair casters	none
Seat compression	Compressed seat height
Footrests	none
Accessories	none

Table 5–4 1988 HFS 100
Environment features and measurements

Feature	Measurement
Illumination	none
Glare	none
Luminance balance	none
Surface gloss	none
Noise	none
Temperature	none
Humidity	none

ANSI/HFES 100 Revision. In 1991, the Human Factors Society initiated revision of the 1988 standard. The revision committee was significantly larger than the original ANSI/HFS 100 committee. It had more than three times the membership of the 1983 committee and represented more organizations (see Table 5–5). The revision committee thus had a broader and more balanced representation than the original committee.

The ANSI/HFS 100 standard revision was necessary because of changes in VDT technologies since its release in 1988 and the necessity for review of new research on VDT ergonomics. The objectives of the revision committee were to:

Table 5–5 ANSI/HFS 100 committee representation

	Original (1988) Committee		Revision Committee	
	Members	%	Members	%
Computer	11	66%	15	15%
Government	1	5%	1	1%
Academic	4	24%	12	12%
Furniture	1	5%	8	8%
Labor	0	%	1	1%
Consultants	0	%	8	8%
TOTAL	17	100%	53	100%

- verify that the requirements and measurements standardized continued to be valid
- change requirements that were no longer valid
- expand the standard to include new technologies (such as flat panel displays and non-keyboard input devices)
- create new requirements necessary because of technology changes or new research
- harmonize the resulting requirements as much as appropriate to ISO 9241, Parts 1–9

Revision Expansion. The proposed revision of the standard resulted in significant expansion of the 1988 version. Its original applications—text processing, data entry, and date inquiry—were expanded to include graphics processing applications, a variety of seated postures, a standing posture, multi-color displays, flat panel displays, and non-keyboard input devices (see Table 5–6). The sections on displays, keyboards, and furniture were also expanded. Whereas the 1988 standard mostly contained requirements for displays and furniture, the proposed revision contains many more keyboard requirements along with descriptions of user performance, biomechanical load, and comfort assessment methods for non-keyboard input devices.

The original version of the standard included four sections totaling 70 pages; the proposed revision may contain over 500 pages. It contains a section on Guiding Principles upon which the design requirements are based. To improve the usability of the previous standard, the revision is presented in a format and style similar to the Mitre guidelines and contains the following with each requirement:

- information supporting or justifying each requirement (in a "discussion" section)
- the identification number of the principle(s) upon which each requirement is based (in a "principle" section)
- references to the research that supports the requirement (in the "reference" section)
- a measurement metric for each mandatory ("shall") requirement (in a "measurement" section)

United States Standards

Table 5–6 Topics in the 1988 and ANSI/HFS 100 standard revision

Topic	1988 Standard	Revision Standard
APPLICATIONS		
Text processing	✓	✓
Data entry	✓	✓
Data inquiry	✓	✓
DISPLAYS		
Monochrome displays	✓	✓
Color displays		✓
CRTs	✓	✓
Flat panel displays		✓
INPUT DEVICES		
Keyboards	✓	✓
Non-keyboard input devices		✓
WORKSTATIONS		
Furniture	✓	✓
Seated posture	✓	✓
Standing posture		✓
Environment	✓	✓
COMPLIANCE MEASURES		
Physical measurement	✓	✓
Performance measurement		✓
Comfort assessment		✓
Biomechanical measurement		✓

ANSI/HFS 100 Conformance. The status of each requirement in ANSI/HFS 100 and its revision (ANSI/HFES 100) is designated by the inclusion of either "shall" or "should" in its statement. As with ISO 9241, "shall" statements are required specifications or practices; "should" statements are recommended. Products and work environments conform to the standard if they meet all of the "shall" mandatory requirements.

Measurement. The 1988 standard specified design principles and quantitative methods to evaluate conformance rather than specific levels of performance, comfort, and biomechanical load. It stated that the quantitative requirements were based on accepted empirical data and established human factor engineering princi-

Table 5-7 Assessment metrics in the 1988 ANSI/HFES VDT standard

Topic	Quantitative Measures
	Number
Displays	11
Keyboards	3
Workstations	3
Environment	0

ples and practices. It specified measures by quantitative metrics and did not include qualitative metrics (see Table 5–7).

The intention of the creators of the 1988 ANSI/HFS standard was that it be used by people with appropriate technical training. Many of the specifications and biomechanical measurement methods require special knowledge and training. The special skills include testing techniques, instrumentation use, analytic techniques, and statistical analysis (see Chapter 9). Examples of the knowledge and skills required include understanding of modular transfer functions and Fourier transforms, use and calibration of photometric instruments and measurement techniques, expertise in conducting visual perception tests, and performing statistical analyses. However, the standard included several specifications that were relatively nontechnical and were able to be easily verified, such as:

- number of blink codes on a display
- visibility of display controls
- accessibility of display controls
- provision of tilt mechanism
- provision of a cursor control
- provision of acoustic feedback
- key top shape
- seat height adjustment mechanism

One of the major contributions of the 1988 standard was the specification of certain measurements that were not previously standardized, such as modulation transfer function—a measure of contrast. This feature is also being incorporated into the proposed revision.

Ergonomic Guidelines

Internal Guidelines. Until the mid-1980s, requirements for software user interfaces were primarily generated on a small scale and within private organizations. One of the first internal software guidelines was published in 1985 by IBM. Its *Guidelines for Man/Display Interfaces* included recommendations for display frame layout, command languages, error prevention and recovery, response times, and behavioral principles. In the same year, Bolt, Beranek and Newman created specification procedures for computer terminal dialogue for the U.S. government.

Military MIS Guidelines. The first software specifications for the military were also in the form of guidelines. In 1982, the U.S. Army published the *Human Engineering Guidelines for Management Information Systems*. These guidelines were intended to foster the inclusion of human factors considerations in the design of Management Information Systems (MIS). Although the guidelines were focused on software, they also included some hardware requirements. For example, Part 1 included software considerations for:

- general design
- dialogue and display development and principles
- language
- file management
- forms, manuals, and microfiche
- training

However, Part 2 addressed hardware features, including:

- keyboard and input devices
- screens and printers
- workstations

Although the MIS guidelines did not include compliance measurement protocols, a checklist was included at the end of each part to determine the status of compliance with the guidelines.

Military Medical Device Guidelines. In 1989, the Department of Defense published *Human Factors Guidelines and Preferred Practices for the Design of Medical Devices.* The standard is being revised and expanded and will include specifications on:

- the human factors engineering process
- human factors and work space (including console design)
- signs, symbols, and markings
- displays, signals, and alarms (including visual displays and audio signals)
- user-computer interface

Mitre Guidelines. In 1984 the Mitre Corporation published *Guidelines on Designing User Interface Software,* the most comprehensive guidelines for text-based software applications. The primary purpose of the Mitre guidelines was to provide a development tool for designers of software user interface. The document contains 994 guidelines for designing software of support the user interface to computer-based information systems. The guidelines cover six areas:

1. data entry
2. data display
3. sequence control
4. user guidance
5. data transmission
6. data presentation

The Guidelines are unique not only because of their extensiveness but also because of their novel format. Each guideline is presented individually and followed by graphic examples of its application, its exceptions, explanations and justifications, supporting research references, and cross-references to other guidelines (see Figure 5–1). The usability of the format as an ergonomic assessment tool was later tested and validated in Hewlett-Packard. The format has been adopted in recent parts of ISO 9241 and in the revision of ANSI/HFES 100.

In addition, many of the specifications in the Mitre guidelines have become a basis for text-based applications of the ANSI/HFES 200 *Human Computer Interaction* standard as well

> **4.3.11 Appropriate Response Time for Error Messages**
>
> Display an error message approximately 2–4 seconds after the user entry in which the error is detected.
>
> EXCEPTION: For type-ahead systems with experienced users, error messages should be displayed as quickly as possible.
>
> COMMENT: Longer delays in error feedback may cause user uncertainty or confusion. Longer delays may also cause frustration if the user is already aware of the error, which is often the case.
>
> REFERENCE: Engle, S.E. and Granda, R.E. (1975)

Figure 5–1 Example of Mitre guideline format (adapted from *Guidelines on Designing User Interface Software,* p. 321)

as the ISO software user interface standards (ISO 9241, Parts 10–17) (see Chapter 3).

Software Vendor Guidelines. Most large software development organizations now have some type of user interface style guide for their own products (see Table 5–8).

Table 5–8 U.S. computer company user interface guidelines

Company	Guideline
Apple Computer, Inc.	*Human Interface Guidelines: The Apple Desktop Interface*
Apple Computer, Inc.	*Macintosh Human Interface Guidelines*
IBM	*A Guide to Software Usability*
IBM	*Common User Access, Advanced Interface Design Reference*
Hewlett-Packard / OSF*	*Motif Interface Guidelines*
Hewlett-Packard	*Software Usability Design Guidelines*
Microsoft	*Windows Interface: An Application Design Guide*
*Open Software Foundation	*OSF/Motif Users Guide*

Each one of these guidelines is created around the appearance and interaction ("look and feel") of a particular user interface style, although they are all supposedly based on human factors engineering (particularly visual and cognitive) principles. There are also marketing agendas integrated into these guidelines. For example, Apple states that the Macintosh guidelines:

- describe an easy way to create products that optimize the interaction between people and the Macintosh computer
- help the guideline user link the philosophy behind the Macintosh interface to the actual implementation of interface elements
- show good human interface design
- provide background information that can help the guideline user plan and make product design decisions

Compared to the Mitre guidelines, the Macintosh guidelines (see Figure 5-2) are visually more appealing because of the extensive use of graphic examples and color. However, according to research, the Mitre guideline format (that is, a one sentence guideline separated from brief supporting information) results in better usability than consolidated information in a paragraph format.

In 1991, IBM published two guidelines on software user interface. One—*A Guide to Software Usability*—describes the process of developing a software user interface. The other—*Common User Access, Advanced Interface Design Reference*—is a composite of software interfaces, conventions, and protocols that provide the framework for the development and production of consistent products. This document combines the best features of the Mitre and Macintosh guidelines except it does not have the beneficial color found in the Macintosh guidelines. It has a format similar to the Mitre guidelines, but provides graphic examples of its recommendations (see Figure 5-3). It is in a "bulleted" format, which has been validated to be more effective than paragraph formats.

Software Standards

ANSI/HFES 200. In 1986, the Human Factors Society formed a task force to investigate the creation of an ANSI/HFS Human

Dialog Box Messages

A good alert box message says what went wrong, why it went wrong, and what the user can do about it. Try to express everything in the user's vocabulary. Figure 11-3 shows an example of an alert box message that provides little information and doesn't suggest to the user what is really going on.

Figure 11-3 A poorly written box message

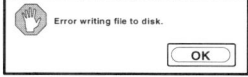

You could improve this message by describing the problem in the user's vocabulary, as show in Figure 11-4.

Figure 11-4 An improved alert box message

Figure 5-2 Example of Macintosh error message guideline (adapted from *Macintosh Human Interface Guidelines,* p. 3)

Computer Interaction (HCI) standard. The original goal of the committee was not to create a standard but to advise the Human Factors Society on software ergonomic activities worldwide and to report on the feasibility of producing an HCI standard. The work of the committee was divided into nine topics:

1. input devices and techniques
2. overview of framework for screen design guidelines/standards
3. windows
4. use of color
5. dialogue techniques
6. user characteristics
7. direct manipulation
8. user guidance
9. evaluation and testing

Action Message

A message that indicates that a condition has occurred that requires a response from the user. The user can correct the condition and then continue, withdraw the request, or get help.

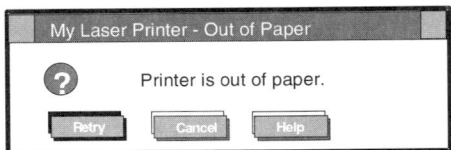

When to Use
- ☑ Display an action message when a situation occurs in which a user must correct the situation and retry, take some related alternative action, or withdraw the request.

Guidelines
- ☐ Provide controls in the message window that allow a user to correct the situation that caused the message to appear or to request a related alternative action...

Essential Related Topics
- Information Message on page 121 . . .

Supplemental Related Topics
- "Audible Feedback" on page 34

Figure 5–3   Example of IBM action message guideline (adapted from *Common User Access, Advanced Interface Design Reference,* pp. 177–178)

The first report created by the HCI committee was a summary of menu-based dialogue principles. Its activities soon expanded into a review of command languages, user assistance, output displays, the design process, and window managers. Over the next few years, most of the activities of the committee were devoted to developing a framework for an HCI standard. It reviewed research on user interface, assembled information, compiled information into guidelines, reviewed software requirements in ISO 9241, Part 1–17 (see Chapter 3), and submitted comments on them to the ISO technical committee responsible for *Software Ergonomics and Man-Machine Dialogue* (see Chapter 1). The technical contributions that HFES-HCI developed and submitted to ISO TC 159 SC4 WG5 include:

- menu dialogues
- command dialogues
- forms-based dialogues
- user guidance windowing systems
- presentation of information
- the design process

In 1994, ANSI sanctioned the committee to create an ANSI/HFES HCI standard. The committee's previous work evolved into the basis of a standard that will include:

- software usability
- software terminology
- menu layouts
- effective use of color
- voice input/output
- command syntax
- graphical user interfaces
- command line interfaces
- special considerations for people with disabilities

ANSI X3V1.9. An ANSI standard is also being developed by the ANSI X3V1.9 committee *User-system Interfaces and Symbols*. The committee's scope is to develop standards for general application areas of text, office and publishing, keyboard and keypads, symbols and icons, and user interfaces. It includes dialog interaction, names of objects and actions, and user guidance. This committee is also revising the existing ANSI keyboard standard X3.154-88.

IEEE, NIST, and OSF. In addition to ANSI, there are a number of other groups in the U.S. that have developed or are developing software ergonomic standards. These include: DOD, DOT and DOE, IEEE, NIST, and OSF.

IEEE is developing standards that facilitate "application and user portability" in *X Windows-based Environments* (P1201). The P1201 committee is divided into two groups:

> P1201.1—The *User Interface Management Systems* group—is focusing its activities on identifying a standard application programmer interface that would allow the creation of both tool kit-independent and presentation-independent applications.
>
> P1201.2—The *User Interface Driveability* group—was creating a "Recommended Practice" document that defined the optimum design for interface components that made a significant effect on drivability —the user's ability to easily transfer from one interface to another. The group developed a template for recording critical information about driveability components, including component definitions, recommended implementation, current industry practice, relevant human factors research and practice, and existing standards that address the components. The draft failed to gain sufficient votes for approval and is currently not active.

The National Institute for Standards and Technology (NIST) (formerly the National Bureau of Standards) standard is based on *X Windows*—the user environment of choice—for networked, bit-mapped, and distributed systems in which application portability is important.

The Open Software Foundation (OSF) is an industrial consortium that, after evaluating a number of *X-Windows* user interface technologies, selected *Motif* as its de facto user interface standard.

Canadian Ergonomic Standards

Guidelines

In 1983, the Treasury Board of the Canadian government released two guidelines on microelectronic technology for federal government users. One was on ergonomic requirements for managers to consider when installing microelectronic equipment; the other addressed health-related concerns associated with the use of VDTs. The information in these guidelines was to be used in evaluating existing equipment, and in replacing old equipment.

Standards

Several private and public organizations also started producing guidelines. In 1989, Canada published the *Office Ergonomics* standard CAN/CSA-Z412-M89. The standard was developed for designers and managers of office environments. Requirements of the office ergonomic standards are located in seven sections:

1. office equipment
2. office furniture
3. visual environment
4. acoustical environment
5. thermal environment
6. air quality
7. ergonomic analysis

It also contains a section on how to conduct ergonomic analyses and includes metrics, such as:

- task descriptions
- worksite dimensions
- checklist of problems and solutions
- link analyses

Unlike the membership of ANSI/HFS 100, the committee members that developed the Canadian standard were not from the computer industry (see Table 5–9). They were mostly representatives of government organizations.

The committee composite may account for the difference in philosophy and content between the U.S. and Canadian ergonomics standards. Whereas the U.S. ANSI/HFS 100 standard mainly contained requirements for *equipment* (such as displays and chairs), the Canadian standard is composed of requirements for the office *environment*.

Another major difference between the two national standards is the amount of background information presented with each requirement. The Canadian standard provides an opportunity for the reader to become familiar with the justification and rationalization for a requirement before it is presented. For example, the functions of the visual, auditory, and musculoskeletal

Table 5–9 Canadian ergonomic standard committee membership

Industry/Organization	Number of Members	%
Computer	0	0%
Government	10	33%
University	3	10%
Furniture	1	3%
Labor	5	17%
Industry	9	30%
Health	2	7%
TOTAL	30	100%

system are discussed before the specifications for visual, auditory, and biomechanical aspects of displays and keyboards are presented.

These features make the Canadian standard possibly the most well-designed, understandable, and usable ergonomic standard yet published. The standard's implementation procedures present strategies for incorporating its contents in the analysis, planning, and management of any workplace.

Regulation

British Columbia is the only Canadian province currently creating ergonomic regulation. It has developed a regulation to reduce cumulative trauma disorders (CTDs) to address the following statistics and their large cost to business and government:

- One-third of British Columbia's worker's compensation claims concern ergonomic-related issues and illnesses.
- In the last decade the British Columbia Worker's Compensation Board has paid more than $400 million for over 100,000 claims related to ergonomic problems.

British Columbian is first implementing this regulation into industry sectors with the highest CTD incidence rates before implementing it into other sectors of industry.

The British Columbian regulation is similar to that proposed by CAL OSHA (see Chapter 6). It will apply to a variety of workplaces and includes eleven regulations requiring employers to:

- identify, assess, and eliminate ergonomic risk factors
- develop a compliance plan
- educate and train employees
- annually evaluate the effectiveness of implemented interventions

The CTD regulation is similar to the ANSI Z-365 proposed standard (see Chapter 6) in providing flow charts and checklists that suggest different levels of intervention. Higher checklist scores indicate more complex control measures. The British Colombian checklist includes:

- risk identification
- job analysis
- awkward postures
- working reaches and height
- seating criteria
- lifting tasks
- pushing limits
- tools and equipment handles
- environmental conditions

The regulation also addresses the VDT workstation and includes requirements for work organization, vision, posture, and work environment. Recommendations are also provided for rest breaks, adjustable display screens, and chairs.

Canadian provinces now periodically meet to harmonize ergonomic activities and guidelines. Like the differences in ergonomic focus between other countries and the U.S. (see Chapter 1), there are differences in focus between Quebec and Canada's other provinces. Quebec focuses on social ergonomics and the other provinces focus on human factors engineering.

Summary

Although the North American countries initiated computer ergonomic standards almost a decade after Europe, both the United States and Canada have since become very active in creating standards, in attempting to legislate requirements, and in funding ergonomic research and studies of computer user interfaces. The North American countries have not yet developed any multi-regional legislation (such as the EU display screen directive). This remains a major difference between North America and Europe in the development of VDT ergonomic standards.

Chapter 6

U.S. Ergonomic Requirements for Special Circumstances

Overview

This chapter describes ergonomic standards and related legislation for disabled users and persons suffering from cumulative trauma disorders. It includes a brief description of two U.S. requirements for equipment for the disabled: the *Americans with Disabilities Act* and guidelines instituted by the General Services Administration. It also includes a description of federal and local standards and laws addressing computer devices and repetitive tasks that result in cumulative trauma disorders.

Requirements for the Disabled

Ergonomics standards are intended to bring about designs that accommodate users' needs and capabilities. However, none of the ergonomic standards discussed thus far have addressed product and environmental requirements for disabled users. Since computer products demand certain capabilities of users, the design, installation, and use of these products must allow for the limitations of disabled users.

A disabled person is defined as one who has a *mental* or *physical* impairment that substantially limits one or more major activities, such as walking, seeing, hearing, sitting, standing, lifting, reaching, or working.

A *mental impairment* is any mental or psychological disorder, including mental retardation, organic brain syndrome, emotional or mental illness, and various learning disabilities.

Examples of mental impairments include dyslexia, Down's syndrome, schizophrenia, and manic-depression.

A *physical impairment* is any physiological disorder, condition, cosmetic disfigurement, or anatomical loss affecting neurological, or musculoskeletal systems, or special sense organs.

Examples of physical impairments include deafness, blindness, and loss of motor coordination and strength due to muscular dystrophy, epilepsy, and cerebral palsy.

Physically disabled individuals include a range of people who have:

- movement and strength limitations and must use crutches, braces, canes, or wheelchairs
- musculoskeletal disabilities such as carpal tunnel syndrome, tendinitis, or back disorders
- visual limitations such as partial or complete blindness or color vision deficiencies
- hearing limitations including partial or complete deafness

Federal laws stipulate that the disabled must be provided with products and space that meet their requirements and allow them the opportunity to work. Employers must provide their employees with equipment they can use and facilities in which they can work safely and efficiently. Manufacturers of information technology products must provide their customers and employees with products that meet the requirements for the disabled. In addition, a company's field representatives should be able to show their customers how to install and use products so that they meet the needs of disabled workers.

Since 1989, two major laws have been enacted by the U.S. government to provide equal opportunity in the use of products and facilities by disabled persons:

1. the *Americans with Disabilities Act*
2. the General Services Administration *Public Law for Users with Disabilities*—99-506.

Americans with Disabilities Act

The *Americans with Disabilities Act* (ADA) sets forth requirements for preventing discrimination against persons with disabilities and for ensuring that the disabled enjoy the same opportunities as individuals without disabilities. The ADA consists of five sections, or titles:

Title I prohibits discrimination against the disabled in hiring and employment practices.

Title II prohibits discrimination in public transportation (such as buses and railroads) and requires public transit systems to provide the same level of service to individuals with disabilities as to those without disabilities.

Title III prohibits discrimination against the disabled in public accommodations and commercial facilities. This section mainly addresses removal of physical barriers and discriminatory standards or policies.

Title IV requires interstate and intrastate telecommunications systems to provide telecommunication relay services for individuals with hearing or speech impairments. This section also requires closed-captioning of public service announcements.

Title V contains various other provisions (including Congressional exemption from the provisions of the ADA).

Objective

The intent of the ADA is to eliminate discrimination against individuals with disabilities in hiring practices and in the work place as well as in public transportation, the marketplace, parks, libraries, stadiums, museums, commercial buildings, warehouses, factories, and business offices. Compliance with the ADA is required on the part of:

- private employers with more than fifteen employees
- state and local governments
- employment agencies
- labor unions

Applications

However, the ADA only partially applies or does not apply to businesses with less than fifteen employees, private clubs, airlines, religious organizations, and the federal government, including the U.S. Congress. It does apply to the following kinds of:

- *public facilities,* such as libraries, schools, and exhibition halls
- *private facilities,* such as offices, lobbies, and manufacturing plants

The ADA requirements thus apply to VDTs, peripherals, furniture, and environments in any of the above conditions. A variety of applications for which ergonomic standards have been created will be required to comply with these requirements; they include computer products used for:

- the purchase of tickets in a public facility
- location of reference material by the public in libraries
- school work
- access to information in museums
- product and component assembly and data entry
- automatic money transaction machines used by the public
- telecommunication interaction

Since office products assume the ability of their users to do many tasks, the design, installation, and use of office products must take into account the requirements of all these tasks. ADA requirements also apply to computer desks and chairs in offices. Specifically, usable features for the disabled must be provided in:

- computer screens
- keyboards and other input devices
- printers and other hardcopy devices
- voice systems

Responsibility

The ADA requires that the work environment "reasonably accommodate" the needs of disabled individuals. This includes acquiring or modifying equipment and making facilities accessible and usable to a disabled person. However, an employer does not have to supply an accommodation if it imposes an "undue hardship" on the business. "Undue hardship" means any accommodation that is significantly cost-prohibitive, disruptive, extensive, or harmful to the fundamental nature of the business.

Liability

Violation of the ADA can result in a first-time fine of up to $50,000 and $100,000 for any subsequent violation. In addition, alternative means for dispute resolution may be employed.

 ## Information Technology Access

In 1986, the U.S. congress amended the 1973 Rehabilitation Act to incorporate Section 508, which requires that:

- guidelines be established to ensure that handicapped individuals will be able to use electronic office equipment with or without special peripherals
- agencies comply with these guidelines in the purchase or lease of electronic equipment

Compliance with Section 508 is required of all government agencies. Any bids to obtain government contracts for the purchase of electronic office products must demonstrate that the products comply with this section. Following the passage of Section 508, the General Services Administration (GSA), together with the representatives of the Department of Education, the electronics industry, various other federal agencies, and the disabled community, created a set of guidelines—*Information Technology Access by the Disabled* (99-506)—spelling out requirements under the new law. The guidelines outline management responsibilities and functional specifications required to comply with Section 508. In general, the guidelines require that electronic office equipment, whether purchased or leased by government agencies, be accessible to individuals with disabilities. Accessibility can be provided by the system or by special peripheral equipment.

Managers' Responsibility

The guidelines require that managers ensure that the needs of users with disabilities are identified and incorporated during procurement planning and determination of requirements. Agency procurement officers are required to address the needs of handicapped employees by incorporating appropriate functional specifications as well as any additional specifications determined to be necessary by a particular agency.

Computer Applications

The guidelines apply to stand-alone computers as well as to network systems and mainframes. Specifications are organized according to functional requirements associated with computer input, output, and documentation. Each of these areas must be addressed during product acquisition planning, and procurement.

Input Device Compliance

Input access differs by type and severity of a user's limitations. Some users with disabilities can use a keyboard if it is slightly modified. Others are unable to use a keyboard and require an alternative input device. The guidelines require that computer input be:

- offered with modified standard input devices that provide:
  - multiple keystroke control
  - key repeat rate control
  - input redundancy
  - toggle keystroke control
- available with alternative input devices
- possible with keyboard orientation aids

Key repeat—which occurs when a key is continually depressed—is a problem for users without sufficient motor control. Multiple keystroke control allows an alternative method of operation by enabling a user to depress keys or buttons repeatedly instead of continuously. A key repeat-rate control feature gives a user control over the repeat start time and rate by allowing either the timing parameters to be extended or the repeat function to be turned off.

Users with motor disabilities typically cannot operate common input devices, such as a mouse, joystick or trackball. An input device that has a redundant input mode can provide the functions of these devices through the keyboard.

However, since people with certain motor disabilities cannot use a keyboard, providing an alternative input device is necessary. Examples of alternative input devices include switches, eye-scan controllers, and voice for visually impaired users who cannot clearly see the letters on the keys. Tactile overlays on a keyboard can provide locators. To assist a motor-disabled user, key guard templates with holes corresponding to the location of the keys help the fingers accurately locate keys.

Output Device Compliance

The Section 508 guidelines that apply to computer output include:

- auditory output capability
- information redundancy
- monitor display
  - large print display
  - access to screen image for text
  - access to screen image for graphics
  - cursor presentation
  - color presentation
- documentation that can be converted into auditory or Braille media

Some types of disabilities require a computer system with a speech capability, such as a speech synthesizer. The guidelines specify that this type of input be provided when appropriate, along with adjustable volume and a headset jack. Redundant presentation of output should be provided for certain types of disabled users (such as the vision or hearing impaired); for example, the visual equivalent of auditory information, or vice versa.

In addition, computer systems should, when necessary, have the capability of presenting large-size text on the screen. All default fonts are normally expected to meet the minimum requirements of ISO 9241, Part 3. However, these minimum sizes would probably not be sufficient for many users with low vision. Thus, for the visually impaired, both a large size default font and negative polarity (light characters on a dark background) should be provided.

In addition, the capability of increasing the size of any image on the screen should be provided. Images on the screens that are to be interpreted by the visually impaired should be able to be scanned by Braille.

User selection and setting of color values is also required. System default color palettes should be provided for specific types of color vision problems.

Documentation Compliance

The Section 508 guidelines also require that all documentation relating to system interface should be provided electronically, and in a form that can be used by the disabled.

Design Guidelines for GSA 508

In response to the passage of Section 508, an industry and government cooperative group also developed a set of guidelines on computer design accessibility: *Considerations in the Design of Computers and Operating Systems to Increase their Accessibility to Persons with Disabilities*. These guidelines are divided into recommendations for individuals who have:

- moderate physical impairments
- severe physical impairments
- visual impairments
- blindness
- deafness
- seizures

The guidelines include such requirements as:

- minimal reach distance
- operation without excessive bending or twisting
- low resistance push buttons (less than 0.5 N)
- concave push buttons
- sliding or edge-operated controls

- overlays providing tactile feedback of label differences
- Braille map of controls and brief meanings
- label letters 5 mm in height
- light color text on dark background
- red and aqua (turquoise) or magenta and green indicator lights
- avoidance of red and green lighted displays
- avoidance of highly saturated and high contrast colored lights for displays
- bold fonts for labels
- warning beeps and tones with visual indicators (flashing/changing color)
- beeping tones below 750 Hz

Ergonomic Health Requirements

The Rise of Cumulative Trauma Disorders

Cumulative trauma disorders (CTDs)—sometimes referred to as repetitive strain disorders (RSIs)—have become the fastest-growing work injury/disorder in the U.S. over the past decade (see Figure 6–1). They now account for over 60 percent of worker compensation claims. (Federal officials attribute part of this rise to improvements in record keeping.) In 1992, the American Academy of Orthopedic Surgeons estimated that in 1984 RSIs cost the U.S. $27 million in lost earnings and medical expenses.

In spite of the fact that there is disagreement in the medical community regarding treatment and prevention, there is now general agreement that a number of factors significantly raise the risk of CTDs. These include:

- repetition
- force
- vibration

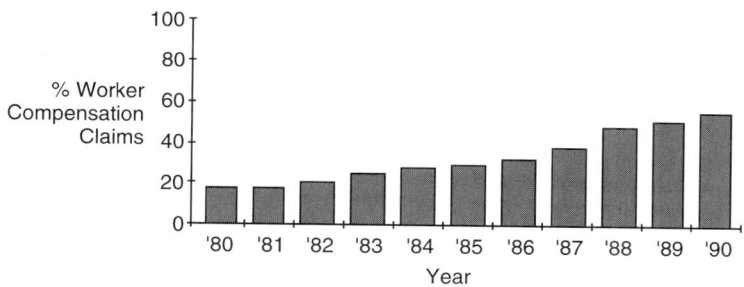

Figure 6-1 Rise in CTDs in the United States in the 1980s

+ temperature
+ posture deviation from neutral

Two of these factors, repetition and postural deviation from neutral, are common in users of computer input devices, such as keyboards, mice, pucks, lightpens, and trackballs. There has been a high rate of CTDs among keyboard operators for at least a decade and recently there has been a growing number of CTDs reported among users of the other input devices.

California Safety Standards

The CAL OHSA VDT Study

In 1987, CAL OHSA (California Occupational Health and Safety Administration) appointed an advisory committee to study the need for VDT-related standards. The committee included ergonomists, ophthalmologists, an optometrist, labor representatives, a psychiatrist, industry representatives, industrial hygienists, and an occupational physician.

Over a two-year period, the committee evaluated ergonomic and medically published studies and research on the following problem areas that appeared to be associated with use of VDTs:

1. vision
2. musculoskeletal problems
3. stress
4. pregnancy
5. indoor environment

In 1989, CAL OSHA published the findings and recommendations of the committee. The committee reported the following general findings:

- The most frequent complaints from VDT users were associated with vision. However, no conclusive evidence was found of permanent damage to vision from using VDTs.
- Extended VDT work was associated with increased rates of musculoskeletal discomfort and disorders; these disorders could be significantly reduced by equipment and job redesign. [Research has since shown that computer input redesign can also reduce CTD risk.]
- Features of VDT work can be associated with stress.
- Although there were significant user concerns regarding adverse pregnancy outcomes associated with VDT work, no conclusive evidence was found as to whether or not there was an association with VDTs.

Since there was no general consensus among the committee members regarding recommendations, industry representatives independently reported the following conclusions:

- VDT regulations are not appropriate.
- Rest breaks are inappropriate.
- A new training program is unnecessary.
- Visual exams for VDT users should be paid by employees.
- There are no causal relationships between VDT design and health disorders.

At the same time, the health and labor representatives on the committee recommended the following:

California Safety Standards

- the establishment of VDT regulations
- the use of existing standards (such as ANSI/HFS 100) as a basis for regulations
- the requirement of worker training on VDT issues
- the regulation of VDT work breaks
- the prohibition of computer surveillance
- additional research on all issues

The CAL OSHA study was submitted to the state government and subsequently became the basis of proposed state and county legislation.

San Francisco VDT Ordinance

In 1992, the San Francisco Board of Supervisors responded to a recommendation of the California Division of Occupational Safety and Health (CAL OHSA) by proposing an ordinance to help reduce cumulative trauma disorders in the workplace. The Board of Supervisors developed the ordinance because of its belief that cumulative trauma needed to be controlled in order to reduce injuries and their resulting cost to employers. This ordinance would have required:

- employers of more than fifteen workers to determine CTD risk
- specific reporting procedures
- worksite evaluations
- implementation of cumulative trauma disorder risk controls
- training of employees in the proper use of equipment and the risk of inappropriate use

Considerable opposition arose, particularly on the part of the business community, because of the presumed high cost of implementing the ordinance and the lack of agreement in the medical community regarding the cause and control of CTDs. The Court of Appeals upheld a Municipal order to nullify the San Francisco ordinance in 1993.

CAL OHSA CTD Bill

In a subsequent effort to address the findings of the CAL OHSA advisory committee and to control CTDs and reduce the increasing cost of CTDs, the California Occupational Safety and Health Standards Board (CAL OSHA) attempted to pass a regulation: *California Code of Regulations: Occupational Noise and Ergonomic Hazards—Prevention of Cumulative Trauma Disorders (Ergonomics)*.

The CAL OHSA regulation would have established minimum requirements for controlling the risk of developing cumulative trauma disorders. It would have applied to all types of repetitive work that included VDT use (except construction). It specified procedures for gathering information on CTD risks, worksite evaluations, control measures, medical management, and training.

Specifically, the CAL OSHA ordinance would have required the following:

- Preliminary information gathering—Employees would have been encouraged to report CTD risks or symptoms, without fear or reprisal or discrimination. Employers would have been required to examine the past three years' medical, safety, and worker's compensation records for evidence of CTD symptoms or risks.
- Work-site evaluation—Employers would have been required to evaluate work activities to assess whether there was in fact a risk if an employee reported CTD symptoms or risks. If a risk was found, modifications to work stations and tools and work environment would have been required.
- Controls—Employers would have been required to modify equipment or work tasks to eliminate or reduce health risks. For employees who performed routine keyboard work for

four hours or more each day, adjustable workstations (tables and chairs), wrist and footrests, proper lighting, and a break every two hours would have been required.

- Medical Management—Employers would have been required to provide free medical evaluations when feasible and to implement measures necessary to prevent injury or the aggravation of symptoms of CTDs. These measures could have included redesign of work duties to reduce CTD risk and the provision of protective equipment like supports, padding, and splints (although splints and palm rests have since been shown to exacerbate CTD problems).
- Training—All employees would have been required to receive training in order to become more aware of symptoms and the impact of CTD and CTD risk factors, the importance of reporting early symptoms, and their right to a free medical evaluation. Employees would also have received information on proper adjustment and use of workstation and equipment and work breaks. Employees would have been consulted for ideas that might result in the reduction of CTD risks.

Representatives of some of California's largest business groups opposed the standard because of its projected implementation and maintenance costs. They also stated that they were not convinced that the medical community agreed on the causes, treatment, and prevention of CTDs, and were opposed to fines (which would have ranged from $7,000 to $70,000). The bill has been redrafted and distributed for approval. The state requires that a bill addressing CTDs be passed by January 1, 1996.

National Safety Ergonomic Standards

In 1992, the U.S. Federal Occupation Safety and Health Administration published an advance notice for proposed rule making to create an ergonomic standard for musculoskeletal disorders. A short time later, the House of Representatives passed a bill that would require OSHA to establish a generic standard on ergonomic hazards by January 1995.

The purpose of the OSHA standard would be to:

- prevent occurrence of work-related musculoskeletal disorders
- inform all employees about musculoskeletal disorders and the risk factors that can cause or aggravate them
- promote continuous improvement in workplace ergonomic protection
- encourage new technology and innovation in ergonomic protection
- identify design principles that prevent exposure to risk factors
- ensure ongoing and consistent management leadership and employee involvement in dealing with ergonomic issues.

The standard would apply to all employers in every type of industry. In addition, it would apply to all parts of the body and thus would not be focused solely on upper limb disorders or keyboarding. Like the defeated San Francisco VDT bill, the OSHA standard would require employer reporting procedures, employee training, risk assessments, record examination, job improvement, and medical management.

Currently, OSHA's only power is to fine employers for ergonomic risks under the general duty clause of the OSHA law, which is a general requirement for a safe workplace.

ANSI Z-365 Standard

In response to the CTD problem, the National Safety Council, accredited by ANSI, appointed a committee to draft a technical standard: *Control of Work-related Cumulative Trauma Disorders.* The committee consists of representatives of universities, medical institutes, insurance companies, industry, professional societies, labor unions, and government agencies.

National Safety Ergonomic Standards

The proposed standard contains principles and practices that the committee believes are necessary for an effective program for controlling CTDs. Although the current document focuses primarily on the upper extremities, its provisions are to be extended to other areas of the body. The standard contains requirements and information on:

- work surveillance
- job analysis and design
- medical management
- training

The purpose of the standard is to control work-related trauma disorders resulting from stress to muscles, nerves, tendons, and associated soft tissues of the body that may occur during work activities and in using work equipment. The standard prescribes assessment methods and requirements for:

- work postures
- work layout
- strength
- vibration
- work rates
- tool design
- workstations

The assessment methods include questionnaires that ask workers to identify areas of pain or discomfort on their bodies, the characteristics of the problem, duration, and general information about their jobs. Some of the comfort assessment scales are based on those researched and validated by European researchers.

The proposed standard includes a risk assessment tool in the form of a checklist and scoring system, which indicates the complexity of corrective measures. Scores are determined by worker motion, posture, velocity of movement, repetition rate, task duration, and force.

As a technical standard, it is intended to be used by persons who have responsibility for the design and operation of work equipment and procedures and for the management of medical, health, and safety programs. The principles and practices are stated in general terms, and professional judgment is required to apply them to specific work situations.

Summary

Ergonomic standards, laws, and guidelines now address the needs of workers with disabilities in the U.S. There are laws relating to the work environment and to equipment that is used by the disabled. In addition, federal, state and local governments are in the process of creating laws and standards for tasks and workstations in which repetitive work is involved.

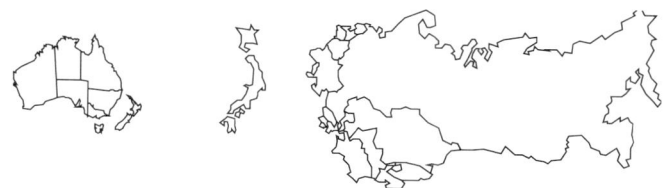

Chapter 7
Ergonomic Standards in Other Countries

Overview

This chapter provides a general overview of some of the ergonomic standards in Australia, Japan and the former Soviet Republics. Australia and Russia are making significant contributions to ergonomics in the field of biomechanics; Japan's contribution is in the area of vision, particularly viewing VDTs. The research of these countries is having an important impact on the application of ergonomic principles to design, on ergonomic standards, and thus on the global economy.

Australia

The RSI Epidemic

At the same time that Europe was creating ergonomic health and safety standards, Australia began creating ergonomic requirements for computer products. Unlike Europe, which was primarily creating VDT standards because of reports of discomfort at VDT workstations and health concerns, Australia was responding to an almost epidemic number of VDT users with musculoskeletal disorders. In the early 1980s, a reported 34 percent of Australia's telecommunication VDT workers had repetitive strain injuries (RSIs). The rate of reported cases of RSIs in other countries did not come close to this figure. However, in the late 1980s, other countries began to show a high incidence of musculoskeletal problems (see Figure 7-1).

RSI Liability

In Australia, injuries sustained as a result of repetitive strain by employees and others using information technology equipment have the same legal consequences as other types of work-related injuries. Liability of employers and suppliers for loss or damage because of RSI is governed both by common law and by State and Commonwealth legislation.

The basis of common law liability for loss or damage from RSI suffered by employees or users is the tort of negligence. In Australia, an employer has a legal obligation to safeguard employees from injury. A computer supplier has the same responsibility to users of equipment supplied by a vendor. Factors that are considered in liability cases include

- training in the use of equipment
- efficient and ergonomic design of workstation equipment
- regular medical exams of employees
- medical monitoring of employees for work-related problems
- sensible and varied workload

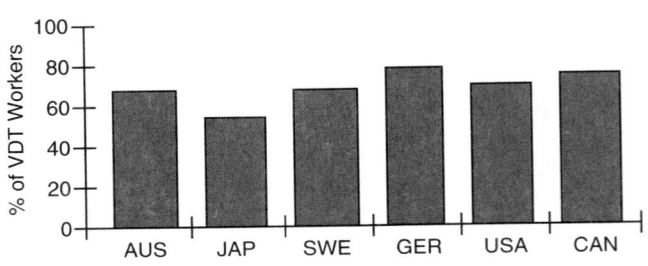

Figure 7-1 VDT users with musculoskeletal problems (from Baummer, 1988)

Australian courts are permitted to use ergonomic standards as a guide in determining the liability of an employer or supplier.

Screen-Based Guidelines

The first publication that addressed the VDT situation in Australia was the *Guidelines for the Introduction and Use of Screen Based Equipment*. These guidelines were the result of an agreement between the government and unions. They were developed and published by the District Public Works in 1984.

The guidelines established policies for the introduction and installation of screen based equipment into the Metropolitan Board of Works—a composite of municipal officers, engineers, scientists, drafters, supervisors, and technical professionals. The guidelines consisted of five basic requirements:

1. All office equipment and associated furniture introduced into the Board were required to have sound ergonomic design.
2. The design of the work environment was required to be in accord with accepted ergonomic principles.
3. All operators of screen based equipment were required to have a medical eye test prior to being trained in the use of the equipment and at regular intervals thereafter.
4. Operators were required to be properly trained. The training was required to include appropriate operating positions and postures as well as recommended work practices and procedures to avoid potential medical complaints.
5. All future proposals to acquire and/or install modern technology equipment in the office environment were required to be subject to an ergonomic assessment.

The guidelines consisted of two sections: (1) procedures and (2) specifications.

The procedures described requirements for

- consulting
- work design
- health surveillance
- product maintenance

Before the purchase of screen based equipment, an ergonomic inspection of the ergonomic appropriateness of the equipment was to be conducted by a qualified professional. These inspectors were required to assess efforts made to meet the requirements of the guidelines. The guidelines also required the replacement of any existing screen-based equipment that did not meet its requirements. The specifications section of the guidelines contained requirements for screen based equipment, workstations, and environments (see Table 7–1).

RSI Regulation

Australia's first legislative response to the RSI problem was to publish a 1986 interim safety regulation—*The Prevention and Management of Occupational Overuse Syndrome*—regarding VDT use. This safety regulation was published by the National Occupational Health and Safety Commission. The regulation contained

Table 7–1 Guideline topics

Screen based equipment
   general design of equipment, displays, & keyboards
   display images
   keyboard layout & profile
   safety
   radiation
   software

Workstations
   chairs
   work stations
   foot rests
   document holders
   documents

Environment
   lighting
   room climate
   noise
   decor
   working space
   safety

requirements for employers and was intended to prevent RSIs by modifying the work environment. The regulation specified requirements for

- work organization
- work rates
- peak demands
- machine pacing
- bonus/incentive schemes
- electronic monitoring supervision
- task design
- task variation/work pauses
- work adjustment periods
- workplace and environment design
- technology selection
- equipment and machine design
- training
- job design

It required that all adjustment mechanisms be tested for

- ease of use
- access
- durability
- safety

Recommended test methods included measuring

- height adjustment
- reach dimensions
- tilt

Products were required to reflect the physical needs of users in respect to:

- anthropometric and kinetic characteristics
- tactile feedback
- visual characteristics
- software
- layout
- movement analysis

Keyboard Regulation

The safety regulation also suggested that further guidance was recommended. Australia met this suggestion in 1988 by publishing a Code of Practice: *Guidance Note for the Prevention of Occu-*

pational Overuse Syndrome in Keyboard Employment. This regulation was more focused than the 1986 regulation because it specifically addressed and described requirements for keyboard equipment and for a wide variety of tasks in which keyboards were used including data processing, banking, telephone operations, programming, and computer-aided design.

The 1988 regulation continues to be the most comprehensive yet published for keyboard use. It expanded the 1986 regulation by adding requirements for each one of the regulatory areas. For example, it specified optimum and maximum reach from a keyboard and the location of documents for viewing during keyboarding.

Workstation Standard

In addition to the 1988 Code of Practice, Australia published the standard *Screen Based Workstations* in 1988. This standard was intended to complement the 1988 regulation.

Like ISO 9241, the 1988 workstations standard is performance based; that is, it specifies performance requirements rather than design requirements whenever possible. For example, it includes a performance test for legibility rather than requirements for dot matrix size and font styles for computer screens. The standard contains two parts:

Part 1: *Visual display units and input devices*
Part 2: *Workstations*

Part 1 includes requirements for VDU construction, adjustability, characters, characteristics, noise, and cleaning. Part 2 includes requirements for keyboard and other input devices. The standard also describes measurement methods for

- surface reflectance
- legibility
- character dimensions
- display luminance
- pixel size
- geometric stability (jitter)

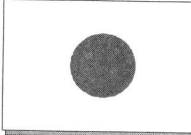

Japan

In general, the Japanese philosophic approach to ergonomics is aimed at improving the quality of life through personal or emotional enrichment. Since the 1980s, Japan has appeared to expand and shift its research focus from engineering advances in function, reliability, and cost reduction to improving its products in terms of comfort, enjoyment and usability. The Japanese believe this shift in research and development is improving their global competition.

MITI Consortium

For several years, Japan has been conducting task forces on improving the ergonomics of their products. Representatives from major companies meet one week each year to determine policies and practices that will make products easier to use and optimize the performance, comfort, and health of users. They meet every two months to review the usability of each other's products and exchange ergonomic knowledge and data base information. All the companies are members of a consortium started in 1987 and sponsored by the Ministry of International Trade and Industry (MITI). The primary goal of the consortium is to enhance user interface (which they define as "ease of operation, reading, understanding, and appeal") with electronic products and environments. Other goals are to decrease product size and enhance user friendliness (which includes product appearance).

VDT Standards

In 1985, the Japan Industrial Safety and Health Association published *Guidelines to Occupational Health in VDT Operation*. These guidelines include requirements for hourly rest breaks, proper lighting and glare prevention measures, physical examinations of workers, and adjustable equipment (see Table 7–2). The guidelines are not legally binding but are recommended standards. They apply to the occupational health of employees engaged in such office tasks as

- input, retrieval, and verification of data
- preparation, editing, and revision of documents
- programming

Ergonomic Standards

Japan's first national ergonomic standard was: *Ergonomics—Anthropometric and Biomechanic Measurement* (JIS Z 8500). The Japan Ergonomics Research Society (JERS) was subsequently authorized to develop ergonomic standards for computers and associated tasks. The JERS developed the following two standards, which respectively correspond to ISO standards 10075 and 9241, Part 3:

1. *Ergonomic Principles Related to Mental Workload—General Terms and Definitions*
2. *Ergonomic Requirements for Office Work with Visual Display Terminals (VDTs)—Visual Display Terminals (JIS Z 8513)*

Adapting to ISO 9241

Japan has a number of problems applying ISO 9241, Part 3, to its display products. One problem is that the recommendations and re-

Table 7-2 Japan's guideline topics and requirements

Environment
   Lighting
   Glare prevention
   Noise control
   Other
Work practice
   Operation hours
   VDT equipment
   VDT furniture
   Adjustments

quirements of ISO 9241-3 focus on Latin, Cyrillic, and Greek alphabetic characters and Arabic numerals. It excludes any ideograph characters such as the Japanese Kanji, and CJK (Chinese/Japanese/Korean). In addition, ISO 9241-3 only considers horizontal reading and writing from left to right, which does not accommodate Japanese reading and writing. To address these problems, the Japanese VDT standard will be a direct translation of ISO 9241-3 but will include Kanji, Katakana, and Hiragana considerations and vertical top to bottom image display characteristics.

CRT—Keyboard Standard

Japan has also created a standard on CRTs and keyboards—*CRT Display and Keyboard Units for Business Use* (JIS 6041)—which was published in 1987. This standard specifies requirements for monochrome and multicolor CRTs and for keyboards that are to be used for a variety of tasks, such as

- data entry
- collation and retrieval of data
- document processing

This standard solved the problems of applying ISO 9241, Part 3, to the Japanese language by separating its requirements into specifications for Kanji and for alphanumeric and Katakana (see Table 7–3).

Table 7–3 Examples of CRT specifications in JIS 6041

Feature	Languages	
	Kanji	Alphanumeric and Katakana
Dot matrix	$\geq 15 \times 16$ dots	$\geq 5 \times 7$ dots
Character height	≥ 3.6 mm	≥ 2.6 mm
Character width	$(0.8–1.2) \times$ height	$(0.5–0.8) \times$ height
Space between characters	$(0.1–0.5) \times$ height	$(0.2–1.0) \times$ height
Space between lines	$(0.2–1.0) \times$ height	$(0.6–1.5) \times$ height

Display requirements include specifications for:

- symbol size
- cursor distinguishability
- luminance for positive and negative polarity
- contrast
- image distortion for nonlinear and raster distortion
- resolution
- display color
- flicker
- glare
- x-ray emissions

The keyboard requirements include specifications for:

- key force
- key stroke
- key shape
- key top dimensions
- key pitch
- key layout
- feedback

The standard also includes requirements for the construction and appearance of displays and keyboards, specifying:

- display adjustment
- keyboard slope
- keyboard stability
- keyboard height
- keyboard surface gloss

A primary difference between Japanese VDT standards and ergonomic standards from other countries is the amount of technical detail included in descriptions of measurement equipment and procedures. For example, the measurement section of the Japanese standard is the most extensive of any ergonomic standard yet published. It describes not only compliance metrics but also provides graphic depiction of measuring instruments and their circuitry.

Flicker Standard

In 1988, the Japanese standards association created an additional standard—*Measuring Methods of Phosphor Persistence for CRT Screens*—to measure display screen flicker. It contains a variety of input measurement methods, including:

- light output time
- pulsed spot method
- integrated light method
- pulsed rather method
- ripple ratio method

Japanese Ergonomic Research

Two research programs have exemplified Japan's shift to ergonomics. One is the Human Technology Project. The other is a national research and development program—the Human Sensory Measurement Applications Technology project. It was initiated to develop technology and data that support product design that reflects user performance, efficiency, and sensory factors (such as familiarity and comfort). The research and development for this project will be conducted by the Research Institute of Human Engineering for Quality of Life. Its purpose is to develop

- non-invasive technologies to measure physiological influences from external stimuli
- simulated-environment technology that can generate and control heat, noise, light, and other external stimuli
- correlation analysis techniques for external stimuli, physiological responses, and sensory reactions

Former Soviet Republics

The Former Soviet Republics have made significant research contributions to the field or ergonomics, especially in biomechanics and physiology. They have also made significant contributions in the study of radiation emissions. As a result of their research, they have the strictest requirements for radiation emissions in the world. However, they have few ergonomic standards for other product features, product usability, and comliance testing.

However, several Ukraine organizations have recently developed a standard to evaluate the quality of VDTs and their environments. The standard is based on the Swedish TCO guidelines (see Chapter 4). However, the only physical measure required in this standard is for levels of radiation emissions. All other compliance evaluation methods consist of questionnaires intended to determine the psychological and ergonomic quality of VDTs and environments.

Further standardization of product ergonomics is important in this region because its countries often import computer products of inferior quality.

Summary

Ergonomic trends in the Pacific Rim and the former Soviet Republics appear to be strongly influenced by European requirements. As of yet, there are only ergonomic computer standards in these regions. Since these regions represent technology producers as well as large potential markets, companies that wish to market their products there should develop products that meet their standards.

Japanese research projects reflect a motivation to improve product ergonomics and the nation's aggressive competitiveness. Other than the European ESPRIT projects (see Chapter 4), there are no similar projects in other areas of the world. Unfortunately, there are no comparable large-scale ergonomic research projects in the United States.

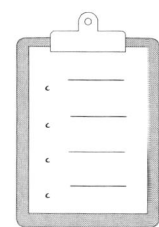

Chapter 8

Ergonomic Checklists

Overview

This chapter provides an overview of the guidelines, standards, and regulations that are expected to dominate ergonomics for the next several years. The material is divided into seven sections:

Part 1 presents a list of minimum safety requirements to which every product should conform. Following the list is a discussion of each requirement.

Part 2 lists tables of all the topics for which requirements exist for each part of ISO 9241 and indicates if each specification is a mandatory requirement or a recommendation.

Part 3 presents a number of minimum hardware requirements that allow conformance to hardware ISO 9241 and most national requirements.

Part 4 lists a number of software user requirements selected from the Mitre guidelines and ISO 9241, Parts 10–12.

Part 5 describes the requirements of the EU display-screen directive. It is included because it is the only international legal requirement that currently exists for computer products and as such has important ramifications for products that are to be marketed in the EU countries as well as associate countries (for example, EFTA countries).

Part 6 presents an overview of the legal requirements of the ADA.

Part 7 presents guidelines for product user interfaces for disabled individuals.

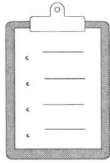

Part 1: Basic Safety and Design Guidelines

Since user safety should be the primary feature of any product, it should be the basic requirement of any product design. The following is a checklist of basic safety requirements derived from the principles of the U.S. *MIL-HDBK-759* (see Chapter 5). The checklist should be used as minimum specifications for all products. A discussion of each guideline is included on the pages that follow.

1. Avoid designs that require user improvisation for operation ❏
2. Design tasks to be definitive and comprehensive. ❏
3. Establish procedures that are as foolproof as possible. ❏
4. Allow corrective actions and provide follow up features. ❏
5. Test product use before releasing it. ❏
6. Design equipment and interfaces to meet user expectations and stereotypes. ❏
7. Do not rely on warning labels or notices to prevent safety problems. ❏
8. Show or describe to users serious results of their actions. ❏

Part 1: Basic Safety and Design Guidelines **165**

9. Equipment should be designed so it is inherently safe to use. ❑
10. Show users results/images of unsafe actions. ❑
11. Consider the possibilities for errors—listing the possible user mistakes and their consequences—and design equipment so incorrect use will do as little harm as possible. ❑
12. Provide complete, legible, and understandable identification of interface components. ❑
13. Every effort should be made to minimize conditions that predispose accidents. ❑
14. Products and interfaces should be designed so that they can be safely used without undue nuisance. ❑
15. Anticipate user mistakes and assure that inevitable mistakes will not injure personnel or damage equipment or components. ❑
16. Understand all the requirements that equipment and interface must satisfy. ❑
17. Imply the sturdiness of a product or interface by its design. ❑
18. Provide detailed information for unskilled users and abbreviated information for skilled users. ❑
19. Create products that are safe and easy to use and maintain. ❑
20. Design interfaces so that use does not depend on assistance from with other users. ❑

Commentary on the Basic Safety and Design Guidelines

1. **Avoid designs that require user improvisation for operation.**
   Comment: If equipment is insufficient or inadequate, users will modify it or improvise to accomplish a task. Improvised equipment leads to improvised procedures and probable hazards.
2. **Design tasks to be definitive and comprehensive.**
   Comment: Some users need little encouragement to discard

standard or recommended procedures and do what they please, which is often the wrong thing to do.

3. **Establish procedures that are as foolproof as possible.**
   Comment: People often feel that their situation is either immune from problems or that problems will occur but to someone else or at some other time. Because people feel safer (or more secure) than they may actually be, procedures must be as foolproof as possible. And since procedures are determined by the equipment (or programs), the equipment (or programs) must be structured to encourage safe use and to minimize the opportunity for unsafe use.

4. **Allow corrective actions and provide follow-up features.**
   Comment: Once a safety problem has arisen, taking corrective action does not necessarily eliminate the problem. The correction may not be sufficient, may not be appropriate, or may even fail to eliminate the problem's cause. If so, the problem may reappear elsewhere or in some other way. Corrective actions should always be followed up to verify that they have really solved the problem.

5. **Test product use before releasing it.**
   Comment: Not all safety problems can be anticipated from studying product design concepts, functional specifications, and engineering drawings. Thorough safety (or interface) analysis requires realistic tests with mockups and prototypes. No matter how simple and foolproof a concept looks on paper, it should be tested before finalizing a design.

6. **Design equipment and interfaces to meet user expectations and stereotypes.**
   Comment: If equipment and interfaces are designed so they do not operate according to the user's expectations and stereotypes, but lead the user to believe that they will, the user will eventually make mistakes.

7. **Do not rely on warning labels or notices to prevent safety problems.**
   Comment: If unsafe possibilities are designed into the equipment (or interface), warnings in a technical manual will not eliminate them. Warning notes have a limited, supplementary value in making mishaps less probable. However, many

users may not read them, may not remember them, or may not know how to find them. Warning notes do not prevent safety problems.

8. **Show or describe to users serious results of their actions.**
   Comment: Accidents seem so remote to some people that they do not appreciate how careless performance can cause accidents. Before users seriously consider and attempt tasks, they must clearly perceive the possible consequences of their actions (including equipment damage or injuries).

9. **Equipment should be designed so it is inherently safe to use.**
   Comment: Equipment should not rely on special user training to prevent accidents. Not all users receive training, even when it is "required." Some users receive outdated training, or intermittent and insufficient training.

10. **Show users results/images of unsafe actions.**
    Comment: Users tend to be unimaginative; they do not visualize the consequences of unsafe acts. They only realize a practice is dangerous after they have seen someone get hurt.

11. **Consider the possibilities for errors—listing the possible user mistakes and their consequences—and design equipment so incorrect use will do as little harm as possible.**
    Comment: Users often use equipment in incorrect and unintended ways. The results of these actions should be anticipated in the design of the equipment, product, and interface.

12. **Provide complete, legible, and understandable identification of interface components.**
    Comment: Accidents often occur because users cannot or do not correctly identify components. Users may injure themselves or damage a component because they do not know what it is.

13. **Every effort should be made to minimize conditions that predispose accidents.**
    Comment: If undesirable conditions are tolerated without a real effort to improve them, they often seem to multiply and interact to become serious safety problems.

14. **Products and interfaces should be designed so that they can be safely used without undue nuisance.**
Comment: If the procedures for safe operation seem needlessly difficult or burdensome, people tend to avoid doing what seems unnecessary. Ideally, equipment should be designed so it is easier to use it safely than unsafely.

15. **Anticipate user mistakes and assure that inevitable mistakes will not injure personnel or damage equipment or components.**
Comment:. Users should be given at least some on-line training with operational equipment; partially trained personnel must be expected to make errors.

16. **Understand all the requirements that the equipment and interface must satisfy.**
Comment: Only by comprehending these requirements can a designer assure that products and interfaces will meet them.

17. **Imply the sturdiness of a product or interface by its design.**
Comment: User care in using and maintaining equipment tends to be related to its complexity and cost. People are most careful of complicated, expensive items. Conversely, users tend to neglect simple, inexpensive items because they seem relatively unimportant.

18. **Provide detailed information for unskilled users and abbreviated information for skilled users.**
Comment: Abbreviated information tends to cause mistakes, but may be useful if users are well trained. Less knowledgeable users should be provided with detailed information. However, as less knowledgeable users may only have access to abbreviated information (like a checklist), it should be as explicit as possible. In general, users tend to experiment or guess at operating procedures, rather than refer to detailed job procedures.

19. **Create products that are safe and easy to use and maintain.**
Comment: Product reputation can be very important because it affects the way it is used and maintained. Even rumors that equipment is difficult or hazardous to use can compound and magnify the basic difficulty.

A designer should remember that most safety problems in equipment arise not from the defects in the equipment but from improper use. Therefore, designers should anticipate how equipment, products, and interfaces might be misused, and design so that misuse will be unlikely and its effects will not be catastrophic. It is obviously more effective to design to prevent misuse, rather than attempt to reduce safety hazards from manufactured equipment.

Ease of use or maintenance affects equipment's reliability. If items are difficult to maintain, users will probably not keep them in good operating condition, so they will not be ready for use when needed. If equipment is difficult to use, operators will substitute other equipment to meet their requirements when they can.

20. **Design interfaces so that use does not depend on assistance from other users.**
    Comment: Equipment is particularly susceptible to misuse if users must obtain assistance from others to use it. People seldom realize that communications are inadequate until they make mistakes, and sometimes not even then

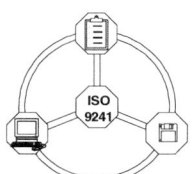

Part 2: ISO 9241 Checklists

The initial table in this section (see Table 8–1) is a summary of the focus, status, compliance measures, and alternative test methods in ISO 9241. Tables (Tables 8–2 to 8–17)[1] list the requirement topics and their conformance requirement: required (mandatory) or recommended. These tables are not provided in the ISO standard but are included here so that they may be copied and used as checklists during a compliance assessment.

(text continues on page 211)

Table 8-1 Compliance Components of ISO 9241

Part	Title	Primary Focus HD	Primary Focus SW	Status	Primary Measure DM	Primary Measure O	Alternative Compliance Test UPT	Alternative Compliance Test CA	Alternative Compliance Test BA
1	General information	✓	✓	IS					
2	Task requirements	✓	✓	IS					
3	Displays	✓		IS	✓		✓		
4	Keyboards	✓		DIS			✓	✓	✓
5	Workstation and posture	✓		DIS	✓				
6	Environment	✓		DIS	✓			✓	
7	Reflections and glare	✓		CD	✓				
8	Color	✓		DIS	✓		✓	✓	
9	Non-keyboard input devices	✓		CD	✓		✓	✓	✓
10	Dialogue principles		✓	DIS		✓			
11	Usability guidance	✓	✓	CD		✓			
12	Presentation of information		✓	CD		✓			
13	User guidance		✓	CD		✓			
14	Menu dialogues		✓	DIS		✓			
15	Command dialogues		✓	CD		✓			
16	Direct manipulation dialogues			CD		✓			
17	Form filling dialogues		✓	CD	✓	✓	✓		

Table Key

Focus:
HW—hardware
SW—software

Measure:
DM—direct measurement
O—observation

Status:
CD—Committee Draft
DIS—Draft International Standard
IS—International Standard

Alternative Test:
UPT—user performance test
CA— comfort assessment
BA—biomechanical assessment

Part 2: ISO 9241 Checklists

Table 8–2   ISO 9241, Part 2: Task Requirements

Requirement Topic	Required	Recommended
Task Design		
Aims		
a) Facilitate task performance		X
b) Safeguard user's health and safety		X
c) Promote user's well-being		X
c) Provide skill development		X
e) Avoidance of undue repetitiveness		X
f) Avoidance of time pressure		X
Characteristics of well designed tasks		
a) Recognition of user experience and capabilities		X
b) Provide for application of user's skills, capabilities, and activities		X
c) Tasks identified as system units		X
d) Contribution of tasks to system function		X
e) Provision of user autonomy		X
f) Provision of meaningful feedback		X

Table 8–3   ISO 9241, Part 3: Displays

Requirement Topic	Required	Recommended
Design viewing disance		
a) Maximum	X	
b) Special applications maximum		X
Line of sight angle		
Angle of view		
a) Minimum angle for legibility		X
b) Not within minimum angle		
1) Specify restricted angle	X	
2) Easy repositioning	X	

(continued)

Table 8–3 *(Continued)*

Requirement Topic	Required	Recommended
Character height	X	
Stroke width	X	
Character width-to-height ratio	X	
Raster modulation and fill factor		
Raster modulation		
a) Monochrome contrast modulation	X	
b) Multi-color contrast modulation	X	
Fill Factor		
Minimum fill factor	X	
Character format		
a) Minimum for numeric and upper case	X	
b) Minimum reading/legibility matrix	X	
c) Matrix with diacritics/descenders	X	
d) Number of pixels for diacritics		X
e) Minimum matrix for super/scripts	X	
f) Minimum matrix for fractions	X	
g) Task unrelated information matrix		X
h) Non-dot-matrix character shapes		X
Character size uniformity	X	
Between-character spacing		
a) Sans serif	X	
b) With serifs	X	
Between-word spacing	X	
Between-line spacing		
a) Minimum spacing	X	
Diacritics/Descenders	X	
Linearity		
a) Maximum row length differential	X	
b) Maximum column length differential	X	
c) Adjacent horizontal displacement	X	
d) Adjacent vertical displacement	X	
Orthogonality		
Horizontal edge-to-mean difference	X	
Vertical edge-to-mean difference	X	
Diagonal edge-to-mean difference	X	
Display luminance		
Minimum luminance capability	X	
Peak luminance	X	
Minimum for luminance coding	X	
Luminance contrast		

Part 2: ISO 9241 Checklists 173

Table 8–3 *(Continued)*

Requirement Topic	Required	Recommended
Minimum contrast capability	X	
Luminance balance		
Frequently sequentially viewed areas	X	
Glare		
Avoidance		X
Additional techniques	X	
Image polarity		
Switchable polarity	X	
Luminance uniformity		
Center-to-edge variation	X	
Within-character luminance variation	X	
Luminance coding		
Minimum luminance coding difference	X	
Blink coding		
a) Attracting attention		
1) Frequency		X
2) Duty cycle		X
b) Readability required		
1) Frequency		X
2) Duty cycle		X
c) Blink control		X
Temporal instability	X	
Spatial instability	X	
Screen image color	X	

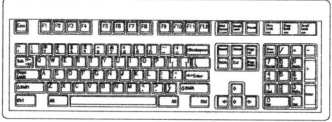

Table 8–4 ISO 9241, DIS Part 4: Keyboards

Requirement topic	Required	Recommended
Dimensions of the keyboard		
Palm rests		
a) Palm rest minimum depth	X	
b) Depth of space below row A		X

(continued)

Table 8–4 *(Continued)*

Requirement Topic	Required	Recommended
Section separation		
a) Perceptual separation	X	
b) Vertical/horizontal key pitch separation		X
Keyboard height		
a) Homerow	X	
b) Row A		X
c) Row B	X	
Slope of keyboard		
a) Home row slope	X	
b) Keyboard slope range	X	
Keyboard profile		X
Keyboard surfaces and material properties		
a) Surface finish	X	
b) Diffuse reflectance	X	
c) Heat conductivity		X
Keyboard placement		
a) Repositioning	X	
b) Detached	X	
c) Stability	X	
Adjusting mechanisms		
a) Slope		X
b) Adjustment mechanism allow stability and placement	X	
c) Adjustment stability	X	
d) Tool free adjustment	X	
Design of keys		
Key layout and center line spacing		
a) Conformance to ISO 9995	X	
b) Horizontal center distance	X	
c) Vertical center distance	X	
d) Non-alpha/numeric center distance		X
Keytop design		
a) Strike surface area	X	
b) Width of strike surface	X	
c) Strike surface for non-alpha and numeric keys		X
d) Surface shape		X
e) Tactile indicators		X
Key displacement and key force		
a) Displacement	X	
b) Displacement for fast keying	X	
c) Ramp action	X	
d) Actuation force	X	

Table 8–4 *(Continued)*

Requirement Topic	Required	Recommended
e) Actuation activation	X	
f) Consistency	X	
Keying feedback		
Key initiated feedback	X	
Preferred tactile feedback		X
a) Visual/auditory with ramp action		X
b) Occurrence of tactile peak	X	
Auditory feedback		X
a) Perceptible	X	
b) Click type		X
c) Volume adjustable	X	
d) Feedback response speed	X	
Visual Feedback		
a) Feedback for long-term status		X
b) Feedback visibility		X
c) Proximity		X
d) Screen indicators		X
Rebound action	X	
Keyboard rollover	X	
Character repeat action		
a) Repeat rate	X	
b) Delay reduction		X
1) User adjustable rate		X
c) Suppression for certain functions	X	
Key Legends		
Character/symbols	X	
a) Conformance with ISO 7000 and 9995	X	
b) Consistency of multiple labels	X	
c) Appearance of character and symbols	X	
Geometric design of Characters		X
a) Legibility	X	
b) Height	X	
1) Full word	X	
c) Width	X	
d) Stroke width ratio		X
e) Contrast	X	
1) Color coding contrast		X
Number and Positioning of Legends		
a) Number of shift levels and legends		X
b) Number of keytop legends		X
c) Alpha and numeric shift levels	X	

(continued)

Table 8-4 *(Continued)*

Requirement Topic	Required	Recommended
d) Positioning conformance to ISO/IEC 9995-1	X	
e) Shift functions	X	
f) Single function legend locations		X
g) Dual function locations		X
h) Overlay function key		X
i) Key proximity		X
j) Provision of reference		X
k) Overlay reference card finish	X	
Durability of legends	X	
Cursors keys		
a) Cursor key provision		
b) Conformance to ISO/IES 9995-5	X	
c) Positioning of "delete" and "erase" keys	X	
Numeric keypad		
a) Zone allocation	X	
b) Telephone layout		X
c) Conformance to ISO/IEC 9995-4	X	
Keytop shape	X	
a) Space bar shape	X	

Table 8-5 ISO 9241, DIS Part 5: Workstation and Posture

Requirement Topic	Required	Recommended
General considerations		
Sitting postures		
a) Clearance for postural changes		X
b) Arm support		X
c) Allowance for postural changes		X
d) Height and tilt adjustment		X
e) Allowance for upright, reclined and and forward posture		X
f) Provision of tilting surfaces		X

Table 8–5 *(Continued)*

Requirement Topic	Required	Recommended
Standing and sit-stand postures		
a) Duration sit-stand		X
b) Alternative with sitting		X
c) Duration of fixed standing		X
d) Leg movement freedom		X
e) Stability	X	
f) Contact points	X	
g) Base	X	
Design reference posture		
a) Thighs horizontal		X
Lower legs vertical		X
Seat height at/below popliteal height		X
b) Upper arm vertical		X
Forearms horizontal		X
c) Hands pronated, no wrist deviation/extension		X
d) Spinal position		X
e) Foot/lower leg position		X
f) Line of sight angle		X
g) Shoulder orientation		X
h) Line of sight		X
Ease of adjustment		
a) Adjustment convenience	X	
b) Operable from usual working position	X	
c) Activation force	X	
d) No training/tools for adjustment	X	
Criteria considered:		
e) System engineering		X
f) Placement of equipment		X
g) Task element location		X
h) Furniture placement in relation to walls		X
i) Ambient conditions		X
j) Placement of additional items		X
Safe controls		X
Clearance under desk by controls		X
Keyboard/display support surfaces		
Clearance under worksurface	X	
a) Clearance for worker and equipment		
b) Comfortable	X	
c) Easy to use	X	
d) Safe	X	
e) Accommodate 5th percentile female and 95th percentile male	X	

(continued)

Table 8–5 *(Continued)*

Requirement Topic	Required	Recommended
Viewing distances and angles		X
a) Screen tilt adjustment		X
b) Screen angle adjustment		X
c) Screen swivel adjustment		X
d) Height adjustment		X
e) Mechanism part of screen/furniture		X
f) Adjustment clarity		X
g) Angle of incidence		X
Finish of work surfaces		
a) Gloss		X
b) Avoidance of high contrast colors		X
c) Reflectance		X
d) Edges and corners		X
e) Radius		X
Safety and stability aspects of workstations		
a) Vibration	X	
b) Stability	X	
c) Unintended collapse	X	
d) Safe adjustment	X	
e) Drawer stability		
f) Drawer contact		
Heat Conductivity limits		X
Chair		
a) Thigh circulation not restricted		X
b) Minimal muscular effort to change posture		X
c) Minimal loading on spine		X
d) Movement freedom		X
e) Medium friction and permeable		X
Seat height		
a) Adjustable height	X	
b) Apply to users	X	
c) Easy adjustment from seated position	X	
d) Stable height	X	
Seat depth		
a) Distance from buttocks		X
b) Adjustability		X
c) Synchronicity		X
Seat angle		
a) Fixed or adjustable		X
Seat width		

Table 8–5 *(Continued)*

Requirement Topic	Required	Recommended
Backrest		
a) Provision for back support		X
b) Lumbar support		X
c) Back rest and seat pan movement		X
d) Back rest height		X
e) Shoulder support		X
f) Forward concavity for high back rest		X
g) Avoidance of excessively curved back rests		X
h) No unintended changes in height		X
Movement of the seat pan and back rest		
a) Independent movement of pan and back rest		X
b) Stability with mechanical lock		X
Arm rests		
a) Height range 5th percentile female to 95th percentile male		X
b) Not restrict preferred posture		X
c) Adjustability		X
d) Height		X
e) Width		X
f) Forward limit		X
Casters		
a) Presence		X
b) Suit to floor and task	X	
c) Frictional resistance while in use	X	
d) Unintentional travel		X
Additional support elements		
Document holders		X
a) Angle and distance adjustment		X
b) Height adjustment		X
c) Accommodates document size		X
d) Non-glossy surface		X
e) Light transmission through document holder		X
f) Stability		X
Footrest		
a) Provision of footrest		X
b) Position		X
c) No unintended movement		X
d) Nonslip surface		X
e) Size to allow movement		X
f) Inclination		X

(continued)

Table 8–5 *(Continued)*

Requirement Topic	Required	Recommended
Support for the hands/wrist		
a) Prevention of static load		X
b) Reduction of deviations		X
c) Support type		X
d) Minimum static posture		X
e) Height and slope match to keyboard		
f) Depth		X
g) Edges		X
h) Width		X
g) Stability		X
Workstations with swivel arms		
a) Use allowance		X
b) Height		X
c) Stability		X
d) Dimensions		X
e) Keyboard positioning		X
Layout of workstation within a room		
a) Access allowance		X
b) Maintenance allowance		X
c) Work flow		X
d) Space allowance		X
e) Shared workstations		X
Cable management		
a) Planning considerations	X	
b) Consideration of wiring and cables	X	
c) Security fasteners		X
d) Length		X
e) Access and maintenance allowance		X
f) Coverage		X

Table 8–6 ISO 9241, DIS Part 6: Environment

Requirement Topic	Required	Recommended
Basic principles for natural and artificial lighting		
Basic aspects		
Harmonization	X	
Flexibility		X
Visibility of screen and documents	X	
Luminance distribution in the field of vision		
Sufficiency		X
Evaluation of psycho-physical aspects		X
Balance		X
Visual communication		X
Safety		X
Luminance ratio balance		X
Glare control		
Avoidance		X
Visual comfort ensured		X
Comfortable posture ensured	X	
Selecting the type of lighting		
Basic aspects		
Selection for ambient light and visual tasks	X	
Sufficient general lighting		X
Individual workstation lighting		
Complementary lighting provided	X	
Lighting direction range		X
Use of color		
Adapted to ambient		X
Recognition of safety codes		X
Low saturation		X
Reflectance gradient		X
Adaptation to visually intense tasks		X
Color rendering and correlated color temperature		
Lamp rendering level		X

(continued)

Table 8–6 *(Continued)*

Requirement Topic	Required	Recommended
Recognition of safety signals	X	
Uniformity with time		
Critical flicker fusion frequency		X
Sound and noise		
Basic aspects		
Reduction of noise effects		
Minimum level		X
Procurement specifications		X
Workroom treatment		X
Mechanical vibrations and electromagnetic fields		
Basic aspects		
Mechanical vibrations		
Basic aspects		
Avoidance of vibrations stress		
Avoidance		X
Reduction		X
Adaptation to individual sensitivity	X	
Workstation shielding	X	
Display image visibility and legibility		X
Component usability		X
Electromagnetic fields and static electricity		
General		
Avoiding adverse environmental effects		X
Thermal environment		
Basic aspects		
Climatic stress		
Air temperature		
Avoidance of heat build-up		X
Protection from thermal stress	X	
Comply with ISO 7730	X	
Comfort zone range		X
Temperature and humidity		
Comply with ISO 7730		X
Workplace and equipment arrangement considerations		
Selection criteria and conformance	X	

Table 8-7 ISO 9241, DIS Part 7: Reflections and Glare

Requirement Topic	Required	Recommended
Recommendations		
Luminance contrast		
a) Maintenance of edge contrast		X
b) Brightness ratio		X
Character luminance contrast relative to specular reflection contrast		
a) Greater than minimum contrast		X
Specular reflection luminance contrast		
a) Less than upper limit		X
Effects of reflection specularity		
a) Absence of task interference		X
Design requirements		
Default color set	X	
a) Compliant ordered pairs	X	
b) ≤ 11 default colors		X
c) Text and background meet 6.5 and 6.7	X	
Color Uniformity		
a) Delta value for adjacent images of same luminance and nominal value		X
b) Delta value for separated colors	X	
Color Misconvergence		
a) 3.4 arc minutes	X	
c) < 20% for + polarity		X
Character height and object size		
a) 20 arm minutes	X	
b) No saturated blue for images < 2 degrees		X
c) Isolated image 30 arc minimum	X	
d) Isolated image 45 arc minimum		X
Chromaticity differences for color sets	X	
Color legibility	X	
Presentation of spectral extreme colors		X
Background and surround image effects		
a) Achromatic background		X
or		
b) Achromatic foreground		X

(continued)

Table 8–7 *(Continued)*

Requirement Topic	Required	Recommended
Chromostereoptic colors not simultaneously presented		X
Number of colors and labels Minimized		
Chromaticity points for color meanings		
Color meaning reference guide		
a) Number of colors for rapid search		X
b) Color differences		

Table 8–8 ISO 9241, DIS Part 8: Color

Requirement Topic	Required	Recommended
Design requirements		
Default color set	X	
a) Compliant ordered pairs	X	
b) ≤ 11 default colors		X
c) Text and background meet 6.5 and 6.7	X	
Color uniformity		
a) Delta value for adjacent images of same luminence and nominal value		
b) Delta value for separated colors	X	
Color misconvergence		
a) 3.4 arc minutes	X	
c) < 20% for + polarity		X
Character height and object size		
a) 20 arm minutes	X	
b) No saturated blue for images < 2 degrees		X
c) Isolated image 30 arc min.	X	
d) Isolated image 45 arc min.		X
Chromaticity differences for color sets	X	
Color legibility	X	

Table 8–8 (Continued)

Requirement Topic	Required	Recommended
Presentation of spectral extreme colors		X
Background and surround image effects		
a) Achromatic background		X
or		
b) Achromatic foreground		X
Chromostereoptic colors not simultaneously presented		X
Number of colors and labels		
Minimized		
Chromaticity points for color meanings		
Color meaning reference guide		
a) Number of colors for rapid search		X
b) Color differences		

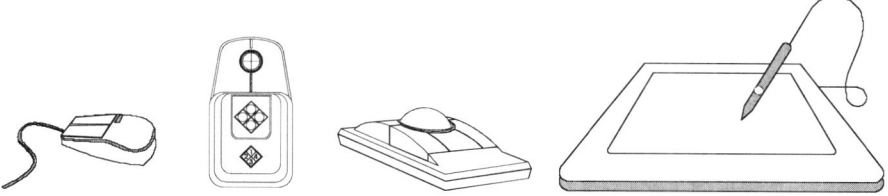

Table 8–9   ISO 9241, CD Part 9: Non-Keyboard Input Devices

Requirement Topic	Required	Recommended
General requirements		
Fine positioning anchor		X
Appropriate resolution provided	X	
Resolution appropriate for task		X
Easy target acquisition and manipulation		X
Repositioning possible without tools	X	
Button design		
Feedback provided	X	
Buttons resistant to inadvertent activation		X

(continued)

Table 8-9 *(Continued)*

Requirement Topic	Required	Recommended
Buttons shaped to assist positioning		X
Force displacement		X
Displacement	X	
Resistance to unintended input	X	
Button lock		X
Ambidexterity		X
Edge characteristics		X
Grasp stability		X
Access from work position	X	
Cable noninterference		X
Expected cursor movements		X
Visual feedback on input	X	
Neutral hand operating posture		X
Accommodation of hand sizes	X	
Provision of hand rest surface		X
Non-obstruction of screen targets		X
Stability		X
Surface temperature		
Lack of parallax between target and device		X
Weight		X
System adjustable gain		X
Button lock provision		X
Specific device requirements		
Mouse and puck		
Location of sensor		X
Ease of moment and resistant to slipping		X
Button perpendicular to force		X
Gain independence		X
Button depression and grip change		X
Lack of reticule parallax		X
Puck window transparency		X
Joystick		
Activation force	X	
Handle size		X
Button location		X
Displacement		X
Trackball		
Chord length		X

Table 8–9 *(Continued)*

Requirement Topic	Required	Recommended
Exposed arc	X	
Rolling force		X
Starting resistance		X
Tablets and overlays		
Contact surface smoothness and flatness		X
Nomenclature and markings		
Size	X	
Width-height ratio	X	
Height-stroke width ratio	X	
Label contrast	X	
Label and background color		X
Force		X
Worksurface height	X	
Tablet depth		X
Label distinguishability		X
Ease of overlay attachment		X
Overlay detachment		X
Overlay flatness		X
Gain		X
Stylus and Lightpen		
Surface coefficient of friction		X
Display selection location		X
Button force		X
Button shape and area		X
Length and diameter	X	
Touch Sensitive Screens		
Target location		X
Touch area		X
Character and symbol legibility	X	
Repeat function delay		X
Inactive space		X
Target spatial and temporal tracking		X
Arm support	X	
Thumbwheel		
Rim exposure	X	
Rim width	X	
Separation	X	
Resistance	X	

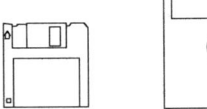
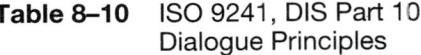

Table 8–10 ISO 9241, DIS Part 10: Dialogue Principles

Section # Requirement Topic	Required	Recommended
Suitability for the task		
Presents information relative to tasks		X
System control non-user tasks		X
Dialogue modified to suit task		X
I/O type and format suits tasks		X
Output presentation matches task requirements		X
Adaptable to recurrent tasks		X
a) Default values provided		X
b) Default values replaceable		X
Original data recall		X
Self-descriptiveness and feedback		
Feedback initiated		X
Severe action consequences explained		X
System model explained		X
Consistent and task familiar terms		X
Feedback at user knowledge level		X
Feedback based on user needs		X
Feedback situation based		X
Minimizes consulting user manual		X
Defaults available		X
Dialogue change notification		X
Controllability		
User controlled interaction speed		X
User configured dialogue		X
a) Dialogue Interrupt		X
b) Continuation		X
Last step undo		X
User control of information exchange		X
User control of data representation		X
User controls amount of data I/O		X
User defined I/O devices		X
Conformity with user expectations		
Consistent behavior and appearance		X
Familiar vocabulary		X

Table 8–10 *(Continued)*

Requirement Topic	Required	Recommended
Similar tasks use similar dialogues		X
a) Standard command structure and syntax		X
b) Consistent dialogues for similar tasks		X
Immediate input display feedback		X
a) Immediate cursor movement		X
b) Minimum cursor movement		X
Response time deviations alert		X
Error tolerance		
Errors explained		X
Special presentation techniques		X
a) User informed of error corrections		X
b) Override option		X
c) Correction possibilities presented		X
Suppression option		X
Additional explanations		X
a) Comprehensible error messages		X
b) Objective error messages		X
c) Constructive error messages		X
d) Consistent error messages		X
User error prevention		X
a) Data verification/validation		X
b) Additional controls for serious consequence commands		X
Error correction w/o switching states		X
Dialogue state change options		X
Consistent and available		
Suitability for individualization		
Adaptation for users w/ different characteristics		X
Choice of representation		X
Modifiable explanations		X
a) Individual naming objects and actions		X
b) Individual functions		X
User control dialogue		X
Select dialogue technique		X
Suitability for learning		
Transparent rules and concepts		X
Task dependent help		X
Support for relevant learning strategies		X
Supports relearning		X
Variety of learning strategies		X

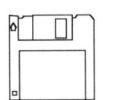 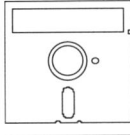

Table 8-11 ISO 9241, CD Part 11: Usability Guidelines

Requirement Topic	Required	Recommended
Specifying and measuring of usability		
Information to be provided		
a) User		X
b) Equipment		X
c) Environments		X
d) Task descriptions		X
e) Interaction specification		X
1) Effectiveness		X
2) Efficiency		X
3) Satisfaction		X
Content of use		X
User description		X
a) Knowledge		X
b) Skill		X
c) Experience		X
d) Education		X
e) Training		X
f) Physical attributes		X
Equipment description		X
a) Hardware characteristics		X
b) Software characteristics		X
c) Materials characteristics		X
Environment description		X
a) Operating system		X
b) Physical environment		X
c) Social/Cultural environment		X
Task description		X
a) Goals/results		X
b) User tasks/activities		X
1) Frequency		X
2) Duration		X
Usability measures		
Choice of measures		
a) Effectiveness		X
b) Efficiency		X
c) Satisfaction		X
d) Effectiveness		X

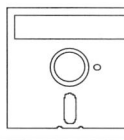

Table 8–12 ISO 9241, CD Part 12: Presentation of Information

Requirement Topic	Required	Recommended
Windows		
Consideration for Multiple Windows		

Criteria for use of windows
  a) Task criteria
  b) System criteria		

Default window parameters		X
Consistent window appearance within an application		X
Indication of primary/secondary window relationships		X
Identification of window control elements		X
Overlapping window format use		X
Tiled window format use		X
User selection of window format		X
Areas		X
Consistent location of areas		X
Provision of unique windows		X
Consistency of window location area		X
Location of control area		X
Input/output area		X
Display of required information		X
Provision of scrolling and paging		X
Display of relative position of display content		X
Groups		X
Sequencing grouping		X
Conventional grouping		X
Grouping by semantics		X
Perceptible distinctiveness		X

Visual chunking
  a) Minimize groups for rapid search
  b) 5 degree size limitation
  c) Maintenance of character size
		X
		X
		X
		X

(continued)

Table 8–12 *(Continued)*

Requirement Topic	Required	Recommended
Lists		
Alignment of vertical lists of alphabetic information		X
Alignment of value or quantities		X
Use of fixed font size		X
Alignment of information with decimal points		X
Organizational order of lists		X
Item numbering		X
Number of lists that exceed space		X
Indication of lists extending beyond display area		X
Tables		
Organization of lists in tables		X
Match between paper and display		X
Continuance of column and row headings		X
Scanning feature		X
Column spacing		X
Labels		
Image labeling		X
Distinguishability of labels		X
Field labels to explain field content		X
Grammar consistency		X
Location of field labels		X
Location of icon labels		X
Location of labels for checkboxes or option buttons		X
Separation of labels		X
Formatting and aligning of labels		X
Inclusion of units of measurement		X
Fields		
Indication of input field format		X
Indication of input field length		X
Distinctiveness of input and output fields		X
Partitioning of length items		X
Cursors and pointers		
Indication of cursor and point position		X
Obscuring of characters by cursor		X

Table 8–12 *(Continued)*

Requirement Topic	Required	Recommended
Non-active location of cursor and pointers		X
Consistency of home position		X
Automatic positioning of cursor at input fields		X
Point designation feature for precision tasks		X
Use of box/block cursor		X
Distinctiveness of cursors and pointers		X
Distinctiveness of active cursor/pointer		X
Distinctiveness of user's cursors		X
Syntactic aspects of alphanumeric coding		
Level of code complexity		X
Distinctiveness of codes		X
Coding text case		X
Use of truncation		X
Use of alphanumeric combinations		X
Grouping of alphanumeric labels		X
Length of codes		X
Application of numeric characters		X
Avoidance of shift characters for codes		X
Evaluation of additional codes		X
Code construction rules		X
Semantic aspects of alphanumeric coding		
Use of standard/conventional meaning for codes		X
Enhancing meaning with associations		X
Consistent use of meaning/function		X
Preference for alphabetic codes		X
Abbreviations		
Minimal length		X
Use of conventional and task-related abbreviations		X
Minimizing deviation from rules		X
Shortening of abbreviations		X
Graphical coding		
Limitation of codes		X
Discernability and discrimination of codes		X
Consistency of code shapes		X

(continued)

Table 8-12 *(Continued)*

Requirement Topic	Required	Recommended
Discriminability of line type		X
Provision for meaning of line orientation		X
Color coding		
Meaningful use of colors		X
Use of colors for multiple meaning		X
Use of color coding conventions		X
Number of colors		X
Avoidance of saturated blue		X
Transfer of meaning to monochrome screen		X
Redundant coding		X
Combining texture with color		X
Avoidance of colors causing chromostereopsis		X
Colors for neutral backgrounds		X
Coding with other visual techniques		
Using special symbols for markers		X
Consistent use of symbols for markers		X
Positioning of markers near associations		X
Use of blink coding		X
Blink coding rate		X
Blinking of additional marker		X
Use of size coding		X
Difference in symbol size		X
Use of brightness coding		X
Use of reverse video		X
Use of underlining		X

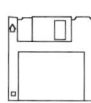 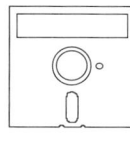

Table 8–13   ISO 9241, CD Part 13: User guidance

Requirement Topic	Required	Recommended
Accessing functions without undue effort or stress		X
Common guidance		
Use of ISO 9241, Part 12, for user guidance		X
Distinguishable from other displayed information		X
Stating results of actions		X
Enhancement of user control perception		X
Removal of non-applicable messages		X
Positive wording		X
Consistent with grammatical construction		X
Construction brevity		X
Natural language voice		X
Information relative to task		X
Use of common terms		X
Neutral wording		X
Non-disruptive messages		X
Reversibility		X
Consistent use		X
Undo		X
Use specified guidance level		X
Prompts		
Indication of type of input		X
Display scenarios		X
Generic presentation		X
Display location		X
Display of default information		X
Provision of cues		X
Automatic cursor positioning		X
To respond to on-line help		X
For current actions		X

(continued)

Table 8–13 *(Continued)*

Requirement Topic	Required	Recommended
Feedback		
For every user input	X	
Nonintrusive, non-disruptive:		
Considerations		X
User skills		X
User variability		X
Task requirements		X
System capabilities		X
Interrupt indication		X
User input acceptance feedback		X
Advisory message of remote service		X
Indication of user request acceptance		X
System response time		X
Highlighting of selected item		X
Status		X
Continuous presentation		X
Automatic presentation		X
Availability on request		X
Consistent window location		X
Visual cue of input disabling		X
On-line help		X
System provision scenario		X
Task specificity		X
Noninstrusiveness		X
User provision scenario		X
Request simplicity and consistency		X
User specification		X
System guidance		X
Requests by nonselection methods		X
Presentation of relevant topics		X
Extent of presentation		X
Predictability		X
Presentation speed		X
Media presentation		X
Modular and layer characteristics		X
Status indication		X
Browse specifications		X

Table 8–13 *(Continued)*

Requirement Topic	Required	Recommended
Navigation		X
Hierarchical structure		X
Self-containment of information presentations		X
Retention of topic upon scrolling		X
User customization		X
Connection with training and documentation		X
Consistency with user needs		X
Provision of context sensitive topics		X
Provision to task information		X
Access to other topics		X
Provision of relevant information		X
Object explanation		X
Visible indication of object explanations		X
Clarification of user ambiguity		X
Error management		
Allowance for continued dialogue		X
Access to additional information		X
System error correction		X
User configuration/indication of system correction		X
Mapping of user inputs		X
Assignment of user inputs with results		X
Avoidance of destructive user actions		X
Assignment of redefined inputs		X
Error correction tools		X
Diagnostic tools		X
Status of system operations		X
User interruption provisions		X
Error explanation		X
Task specificity		X
Discrimination of successive occurrences		X
Removal of error messages		X
Consistency of error message presentation		X
Movement of error messages		X
Presentation order		X
Presentation of alternative inputs		X
User editing erroneous input		X

(continued)

Table 8–13 *(Continued)*

Requirement Topic	Required	Recommended
Multiple error detection		X
User specification of content level		X
a) User control of message presentation		X
b) User control of auditory feedback		X
Indication of potential problem		X
Automatic return to default settings		X
System check of file status upon exit		X
Provision of undo with warning indicator		X
User access to warning information		X

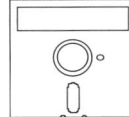

Table 8–14 ISO 9241, DIS Part 14: Menu Dialogues

Requirement Topic	Required	Recommended
Menu Structure		
Structuring into levels and menus		X
Conventional categories		X
Logical categories		X
Arbitrary categories		X
Search time considerations		X
Grouping options within a menu		X
Logical groups		X
Arbitrary Groups		X
Sequencing of options within groups		X
Consistency		X
Importance		X
Conventional order		X
Existing order		X
Frequency of use		X
Alphabetical order		X
Navigation		X
Cues		X
Titles		X

Table 8–14 *(Continued)*

Requirement Topic	Required	Recommended
Numbering schemes		X
Graphic techniques		X
Simultaneous display		X
Menu maps		X
Rapid navigation		X
Hierarchical access		X
Mode access		X
Skipping levels		X
Returning to initial menu		X
Upward level movement		X
Multiple pathways		X
Option Selection and Execution		X
Selection methods		X
Alternative methods		X
Separate actions for selection and execution		X
Fast access		X
a) Bypass mechanisms		X
b) Combining selection and execution		X
Feedback		X
Unselecting options		X
Response delay		X
Alphanumeric keyboard		X
Minimizing keystrokes		X
Command line location		X
Case equivalence		X
Key letter designators		X
Easy rule for designators		X
Number designators		X
Designator structure and syntax		X
Function keys		X
Designators		X
Displaying assignments		X
Menu orientation		X
Consistency of assignment		X
Cursor key selection		X
Options to columns		X

(continued)

Table 8-14 *(Continued)*

Requirement Topic	Required	Recommended
Options in rows		X
Minimizing keystrokes		X
Cursor response time		X
Pointing		X
Pointing area		X
Unintended activation		X
Keyboard equivalence		X
Voice		X
Phonetically distinct		X
Consistency		X
Noise		X
Feedback		X
Deselecting or undoing		X
Menu Presentation		X
Option accessibility and discrimination		X
Critical options		X
Frequent usage		X
Occasional usage		X
Available options		X
Unavailable in addition to available options		X
Selection default/highlighting		X
a) Most frequent option		X
b) First option		X
c) Previous option		X
d) Least destructive option		X
Multiple menus/option groups		X
Multiple selections		X
Explicit designators		X
Implicit designators		X
Placement		X
Consistency of layout		X
Titles		X
Explicit designators		X
Accelerator keys		X
Options in column spacing		X
a) Spacing		X
b) Single spacing		X
c) Option groups		X

Table 8–14 *(Continued)*

Requirement Topic	Required	Recommended
d) Justification		X
e) Multiple columns		X
Options in rows		X
Color		X
a) Group options		X
b) Compliant with Part 8		X
c) < 5 colors in one menu		X
Fonts		X
a) Legibility		X
b) Number		X
Borders or lines		X
a) Simplicity		X
b) Readability interference		X
Textual option structure and syntax		X
Unambiguous names and titles		X
Keywords		X
a) Begin with keywords		X
b) High imagery		X
Option terminology		X
Option phrasing		X
Action options		X
Object options		X
Action and object options		X
Transition to command language		X
Leading to another option		X
Graphic option structure and syntax		X
Icon labels		X
Grouping		X
Visual distinctiveness		X
Voice option structure and syntax		X
Number of options		X
Syntax		X
Acoustic distinction		X
Replay capability		X

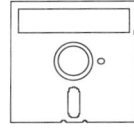

Table 8–15 ISO 9241, CD Part 15: Command Dialogues

Requirement Topic	Required	Recommended
Structure and syntax		
Consistency of command function		X
User creation of macros		X
Combination of multifunctions into single		X
Parameter dependencies and command meaning		X
Syntax appropriateness and consistency		X
Blanks/standard symbol for separators		X
Language familiarity of semantics and syntax		X
Command arguments		X
Clarity of argument modification		X
Keyword use of argument formats		X
Placement cement of optional arguments		X
Separation of arguments		X
Precise and necessary qualifiers		X
Command representation		X
Command names		X
Distinctiveness		X
Use of user familiar language		X
Use of neutral language		X
Command length		X
Suffix addition		X
Abbreviation		X
Type command characteristic		X
Use of simple rules		X
Use of simple truncation		X
Function keys and hot keys		X
Function key consistency		X
Hot key consistency		X
Limitation of modifiers		X
Grouping of modifiers		X
Input and output considerations		X
Reuse of commands without retyping		X

Table 8–15 *(Continued)*

Requirement Topic	Required	Recommended
Capability of keying command series		
Correction for error content		X
Editing of commands prior to execution		X
System acceptance of command misspelling		X
Provision of defaults to minimize keying		X
Provision of undo and confirmation for destructive commands		X
Use of synonyms for commands		X
Echoing of typed commands		X
User control of output		
Format consistency for similar/related output		X
Feedback and help		X
Indication of command acceptance		X
Provision or error feedback		X
Highlighting of area of input error		X
Provision of command information		
Available commands and meaning		X
Appropriate syntax structure		X
Required and optional parameters required		X
Command entry history		X
Provision of performance aids		X
Provision of parameters values and menus for long commands		X

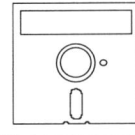

Table 8–16 ISO 9241, CD Part 16: Direct Manipulation Dialogues

Requirement Topic	Required	Recommended
General presentation		
Distinctiveness of objects, attributes, and manipulations		X
Availability of objects, attributes, and manipulations		X
Distinctive states of objects, attributes, and manipulations		X
General interaction		
Pointer access		X
Alternative access		X
Object manipulation syntax		X
Recognizing and pointing		X
Pointing interactions		X
Pointer layout as status indicator		X
Prompting of input options		X
Immediate feedback for stepwise manipulations		X
Continuous feedback for ongoing manipulations		X
Direct manipulation of output		X
Reversibility of manipulations		X
Direct manipulation macros		X
Visual cues indicating system control		X
Metaphor use		
Providing a system framework		X
Recognizability of analogies		X
Analogy to real world manipulations		X
Indicating limits of metaphors		X
Metaphors for object status		X
Realistic feedback to changing object states		X
Metaphors for manipulation feedback		X
Design of object representation		
Differentiating object instances		X
Default content of text labels		X

Table 8–16 *(Continued)*

Requirement Topic	Required	Recommended
Hidden objects		X
Recovery of object states		X
Multiple representation of objects		X
Display of newly created or opened objects		X
Design and manipulation of attributes		
Easy access to attributes		X
Values in property sheets or dialogue boxes		X
Feedback on attribute changes		X
Combined manipulation of attributes		X
Manipulation of objects and object representations		
Generic manipulations		X
Extended selection of objects		X
Simultaneous manipulation of several objects		X
Provision of object handles		X
Automatic prompting of available objects		X
Structuring for object selection and presentation		X
Predefined interactions between objects		X
User control of object positions		X
Access to overlapped objects		X
Access to hidden objects		X
Quick access mechanisms		X
Continuously selected objects		X
Cursor positioning		X
Window manipulation		
Selective changeability of window attributes		X
Manipulating single window position		X
Position of newly created windows		X
Position of restored windows		X
Manipulating visibility of single window		X
Manipulating single window size		X
Manipulating single window scale factor		X
Manipulating presentation state of single window		X
Default position of iconified windows		X

(continued)

Table 8-16 *(Continued)*

Requirement Topic	Required	Recommended
Preventing loss of unsaved work		X
Manipulating activation state of single window		X
Manual assignment of input focus		X
Automatic reassignment of input focus		X
Customization of assignment of input focus		X
Random position of selection indicator in window		
Automatic display of objects in windows		X
Movability of displayed window contents		X
Moving window contents in single units		X
Moving window contents in multiple units		X
User rearrangement of windows contents		X
Manipulating single window panes		X
Feedback on window manipulation		X
Manipulating multiple windows		X
Manipulating visibility of related window groups		X
Manipulating position of related windows		X
Manipulating window attributes with minimal input		X
Consistent behavior of window controls		X
Visual distinctiveness of window controls		X
Appropriate size of window controls		X
Appropriate position of window controls		X
Status indication of window controls		X
Alternative manipulation mechanisms		X
Manipulation of controls		
Indication of availability		X
Easy identification		X
Providing menu bar		X
Selection of menu options		X
Activation of menu bar option		X
Providing pull-down menus		X
Selection of pull-down menu option		X
Activation of pull-down menu option		X
Pull-down menu options applicable to multiple objects		X

Table 8–16 *(Continued)*

Requirement Topic	Required	Recommended
Closing a pull-down menu without option activation		X
Providing pop-up menus		X
Selection of a pop-up menu		X
Selection of pop-up menu option		X
Closing pop-up menu without option activation		X
Pop-up menu options application to multiple objects		X
Providing choice lists		X
Access to choice list options outside the display		X
Selection of choice list option		X
User single selection of mechanisms in choice lists		X
User multiple selection of mechanisms in choice lists		X
Providing drop down lists		X
Display of drop down list options on request		X
Access to drop down list options outside the display		X
Selection of drop down list option		X
Single selection mechanisms in drop down lists		X
Closing drop down list without option selection		X
Providing drop down combination boxes		X
Display of options in drop down combination boxes on request		X
Access to drop down combination box options outside the display		X
Selection of drop down combination box option		X
Single selection mechanisms in drop down combination boxes		X
Closing drop down combination box without option selection		X
Typed entry in drop down combination boxes		X
Filtered display of options in drop down combination boxes		X
Providing check boxes		X
Changing the state of check boxes to on or off		X

(continued)

Table 8–16 *(Continued)*

Requirement Topic	Required	Recommended
Providing groups of radio buttons		X
Changing the state of radio button to on or off		X
Providing pushbuttons		X
Selection and activation of pushbuttons		X
Providing spin buttons		X
Single selection mechanism of spin buttons		X
Types entry in spin buttons		X
Providing sliders		X
Changing values indicated by sliders in single or multiple units		X
Providing scroll bars		X
Providing horizontal scroll bars		X
Providing vertical scroll bars		X
Changing displayed object controlled by scroll bars in single or multiple units		X

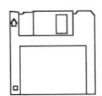
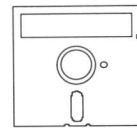

Table 8–17 ISO 9241, CD Part 17: Form-Filling Dialogues

Requirement Topic	Required	Recommended
General		
Titles indicate purpose		X
Discriminable coding		X
Avoidance of density > 40%		X
Provision for forms entering instruction		X
Layout		
Consistency with hardcopy source		X
Grouping on non-document forms by function		X
Positioning of required vs. optional fields		X
Vertical left justification of entry fields		X
Allowable field values		X
Label length variable and justification		X
Left justification for non-variable label length		X

Table 8–17 *(Continued)*

Requirement Topic	Required	Recommended
Column and row labels for multiple fields		X
Page identification		X
Fields		
Display of fixed field length		X
Distinguishability of optional and required entries		X
Description of data entry type		X
Semantics of entry field levels		X
Redundant display symbols/units		X
Display of data entry cues and location of abbreviations		X
Upper case of first letter of field labels		X
Input		
General		
Minimization of cursor movements between fields		X
Direct movement to entry field		X
Provision of default values		X
Minimization of use of different input devices		X
Use of input device for form entry and navigation		X
Text entry		
System justification		X
System entry of leading zeros		X
a) Indication of input area		X
b) Provision of auto wrap		X
Separation of exclusive fields with words		X
Avoidance of interdependency rules		X
Consistency between entry and display space		X
Choice entries		
Viewing and selecting of fill-in options		X
Cue discrimination of choice entries		X
Menus for field entry options		X
Lists for large field/options		X
a) Visual cues for selectable options		X
b) Navigation mode for long lists		X
Buttons for small number of options		X

(continued)

Table 8–17 *(Continued)*

Requirement Topic	Required	Recommended
Changing appearance of buttons for exclusive choices		X
Changing button appearance for state changes		X
Use of stepper button		X
Control		
Undo and back up		X
Navigation to errors		X
Correction of mis-keys		X
Avoidance of entry in restrictive areas		X
Single actions for transmission		X
a) Obvious signal completion and redisplay for new data		X
b) Return to state prior to creation		X
c) Escape provision		X
d) Undo provision		X
Provision of temporary save		X
Feedback		X
Echoing of keyed entry		X
Perception of cursor position		X
Notification of reception of entries		X
Notification of updates to database		X
Notification of immediate error input		X
Navigation		
Automatic cursor position to next entry		X
Interfield cursor placement		X
Undo capability to initial field		X
Tabbing		
Provision of manual tabbing between partially completed fields		X
Provision of auto-skip tabbing for prefilled forms		X
Exclusion of tabbing combinations		X
Choice tab skipping for exclusive fields		X
Intergroup navigation		X
Record cycling		X
Form navigation via pointing device		X
Scrolling		
Forms accessibility		X

Table 8–17 *(Continued)*

Requirement Topic	Required	Recommended
Direct form access		
Form selection via naming/menu selection		X
Intra-form navigation		X
Form navigation in hierarchies		X
Initial form return access		X
Activation of last selected form in window		X

Part 3: Hardware Requirements

The requirements in this section are minimum specifications for displays, keyboards, non-keyboard input device, desks, chairs, and office environments. They are a composite from several national and international standards (Tables 8–18 through 8–22).

(text continues on page 217)

Table 8–18 Display Requirements

Feature	Specification		
	Minimum	Maximum	Preferred
Viewing distance	300 mm	600 mm	400 mm
Viewing angle	5° below horizontal	45° below horizontal	15° below horizontal
Character height	3 mm	4 mm	3.5 mm
Stroke width	1/6 character height	1/12 character height	1/8 character height
Character height to width ratio	0.45 to 1.0	0.9 to 1.0	0.6 to 1.0
Raster modulation			
For monochrome		0.4	
For color		0.7	

(continued)

Table 8–18 *(Continued)*

Feature	Specification		
	Minimum	Maximum	Preferred
Fill factor	0.3		
Character format	5 × 7		7 × 9
Character size uniformity	−5% character height	+5% character height	0 variation
Between character spacing	1 pixel	2 pixels	
Between word spacing	1 character width	1 character width	1 character width
Between line spacing	1 pixel	number of vertical pixels in lower case "i"	
Linearity		2% row/column length	
Orthogonality			
Horizontal edges and mean length		0.02	
Vertical edges and mean length		0.02	
Luminance			
Light characters	35 cd/m^2	150 cd/m^2	80 cd/m^2
Light background	35 cd/m^2	150 cd/m^2	80 cd/m^2
Contrast	3:1	10:1	5:1
Luminance balance	10:1	100:1	10:1
Glare	none	interferes with task	none
Polarity for:			
Text processing			light characters on darker background
Proofing			dark characters on lighter background
CAD			light characters on darker background
Color			light characters on darker background
Bright rooms			dark characters on lighter background
Dark rooms			light characters on darker background
Luminance uniformity		1.7:1	

Table 8–18 *(Continued)*

Feature	Specification		
	Minimum	Maximum	Preferred
Blinking	1 Hz	5 Hz	2 Hz
Flicker	80 Hz	100 Hz	100 Hz
Jitter		0.0002 mm/mm	
Color differences	20 ΔE		40 ΔE

Table 8–19 Keyboard Requirements

Feature	Specification		
	Minimum	Maximum	Preferred
Keyboard profile			dished, stepped, sloped, flat
Key shape			dished and square
Stability	wobble just detectable	wobble interferes with keying	no wobble
Height		30 mm at home row	
Slope	0°	± 15°	−5°
Surface	matte	45 gloss units, silky matte	matte
	dark gray	light gray	dark gray
Color difference	20 ΔE	40 ΔE	40 ΔE
Lightness difference	1:3	1:10	1:5
Numeric keypad			telephone layout
Zero key			below 1, 2, 3, row
Toggle			indicator light
Labels			sans serif font
Single letters	4 mm	7 mm	upper case
Words	2.2 mm	3 mm	upper and lower case

(continued)

Table 8–19 *(Continued)*

Feature	Specification		
	Minimum	Maximum	Preferred
Cursor keys			cross or "T" layout
Key			rollover
Center spacing	18 mm	19 mm	
Top edge spacing	5 mm	7 mm	6 mm
Surface width	12 mm	15 mm	
Force	0.4 N	0.8 N	0.6 N
Displacement	1.3 mm	6.4 mm	2 mm–4 mm
Repeat rate	10/second	20/second	
Feedback			tactile plus acoustic
Visual feedback	0.2 seconds	0.5 seconds	0.2 seconds
Character height to width ratio	50% character height	100% character height	70% character height
Stroke width		1:5	1:6–1:8
Wrist rest	padding		padding and cushion

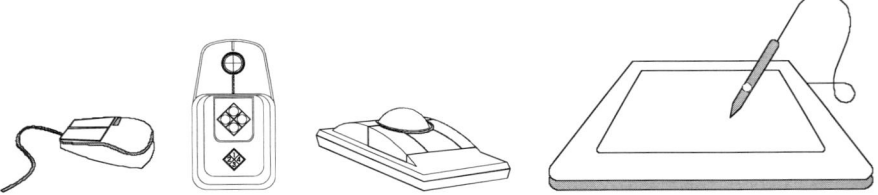

Table 8–20 Non-Keyboard Input Device Requirements

Feature	Specification		
	Minimum	Maximum	Preferred
Mouse			
Sensor			located under fingers
Button motion			coincident to finger motion
Width	40 mm	70 mm	
Length	70 mm	120 mm	
Height	25 mm	40 mm	
Puck			
Reticule			transparent window, no parallax

Table 8–20 *(Continued)*

Feature	Specification		
	Minimum	Maximum	Preferred
Joystick			
Finger force	0.05 N	1.13 N	
Hand force	2 N	5 N	
Finger handle	6.5 mm	16 mm	
Hand grip length	110 mm	130 mm	
Hand grip diameter			28 mm
Angle displacement		45°	
Gain	1.0		
Trackball			
Size	30 mm	100 mm	55 mm
Exposed surface	100°	140°	120°
Starting force	0.4 N	7.0 N	
Resistance	0.3 N	1.0 N	0.3 N
Touch Screen			
Target location		shoulder height	below shoulder
Touch area	10 mm		15 mm
Dead space	3 mm		
Light Pen/Stylus			
Weight	1 oz.	6 oz.	
Surface			low coefficient of friction
Button force	0.4 N	0.8 N	
Length	120 mm	180 mm	
Diameter	7 mm	20 mm	
Shape			round/triangular

Table 8-21 Desk and Chair Requirements

Feature	Specification		
	Minimum	Maximum	Preferred
Desk			
Height	23"	29"	29"
Width	30"		50"
Depth	30"		40"
Knee clearance	16"		20"
Foot clearance	18"		25"
Keying height	27"	29"	27"
Surface			non glare
Chair			
Seat height	13"	20"	
Seat width	18"	20"	
Seat depth	15"	17"	
Back height	15"	25"	
Covering			moisture absorbing
Adjustments	seat height		seat height and angle

Table 8–22 Environment Requirements

Feature	Specification		
	Minimum	Maximum	Preferred
Lighting			
Level	200 lux	700 lux	Natural 500 lux
Type			Indirect
E_z/E_h	0.3	0.5	
Flicker		perceptible not distracting	none
Sound and Noise	30 dB(A)	60 dB(A)	40 dB(A)
Vibrations		perceptible not distracting	none
Electromagnetic Fields	−1 kV/m(0.3)	+1 kV/m(0.3)	
Thermal Conditions			
Heat	68″	72″	70″
Humidity	40%	70%	60%

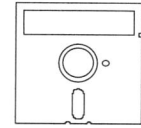

Part 4: Software Guidelines

The following are examples of software guidelines derived from requirements and recommendations in ISO 9241, Parts 10–17. They are presented to provide an overview of requirements and guidelines that address:

1. general interface
2. user expectations
3. user control
4. scrolling
5. error and help messages
6. feedback
7. system individualization
8. keying
9. menus
10. navigation
11. options
12. pointing

1. The general interface should
   - present only concepts and information related to user activities
   - make system perform functions not related to user task
   - make type and format of input and output suit the task
   - match presentation of output with user task(s) requirements
   - provide automatic default inputs
   - provide perceptible indicator for system responses longer than 2–3 seconds
   - allow original data to be quickly and easily accessible in one action
   - allow previous tasks to be immediately accessible

2. The system should conform with user expectations by
   - automatically locating cursor where input is required
   - providing consistent dialogue behavior and appearance
   - providing consistent and available state changes
   - providing vocabulary familiar to user(s)
   - providing similar dialogues for similar tasks
   - minimizing cursor movements
   - allowing for deviations from normal system response time

3. User control should be provided by the software interface and should
   - allow undo of last dialogue step/action
   - accommodate user work speed
   - allow user to control response
   - allow user to determine restart joint of interrupted dilogue
   - allow choice of different interactions levels
   - present input and output data under user control

4. Scrolling features should include
   - vertical scrolling controlled by up and down cursor arrow keys
   - horizontal scrolling controlled by left and right cursor arrow keys
   - movement of one option with each key press

5. <u>Error and help messages</u> should

    - prevent user from making errors, particularly destructive ones
    - provide comprehensive, objective and constructive explanations
    - use non-judgmental semantics
    - help user correct errors
    - help user recognize error situations
    - inform user of automatic error corrections
    - leave correction to user when appropriate
    - provide additional information on request
    - provide input edits to validate/verify data
    - check correctness before activating input
    - provide additional control for commands with serious consequences
    - allow help to be task dependent

6. <u>Feedback</u> should

    - relate to the type of situation
    - be provided after each user action
    - use consistent terminology
    - help user understand system
    - be matched to the user's knowledge
    - be of different types and easily perceptible
    - provide defaults
    - provide information about system status
    - provide information about expected input

7. <u>System individualization</u> should provide an interface that can be adapted to user's:

    - language
    - culture
    - knowledge
    - experience
    - perception
    - sensorimotor skills
    - cognitive abilities

The interface should also allow

- modification of dialogue information to accommodate user knowledge
- incorporation of user's own vocabulary
- addition of functions
- customization of time parameters
- modification of defaults with access to permanent system defaults
- provide different choices of dialogue representation

8. <u>Keying</u> input should be designed so that

- keystrokes are minimized during item selecting and activation
- command line(s) are located at fixed locations
- input is allowed in either type case
- easy to learn key designators are provided
- consistent structure and syntax option designators are provided
- option designators are assigned to correspond to function key labels
- quick and easy access to the file menu is provided
- means for obtaining a menu can be continually displayed
- design menu orientation (horizontal or vertical) is the same as the function key orientation
- the same function key can be used consistently to select and execute function keys

9. <u>Menus</u> should

- use option names most representative of option function
- use fonts with good legibility at a variety of reading distances
- use keywords with strong cognitive connection to action or object
- use terminology familiar to users
- state options concisely
- use consistent option phrases
- use verbs for option names that represent actions
- use nouns for option names that represent objects

- use verb-noun syntax for options that represent both actions and objects
- provide appropriate cues when options lead to another menu of other kind of dialogue
- make option names and group titles semantically distinct
- make option names begin with the word most representative of the option

Menus should be designed so that

- options of variable length are relative to each other
- menu options of fixed length are in the same position in each menu
- menu titles are at the top of the menu panel
- explicit designators are to left of option name
- option accelerators or shortcut keys are to the right of option name

Menus should be structured so that

- menus can be formatted into a consistent structure
- options can be placed into groups
- user expectations are reflected
- users are aided in finding and selecting options
- options are arranged into conventional/natural groups familiar to users
- options are arranged into conventional/natural groups easily learned
- options are grouped to minimize levels and maximize number of options
- scrollable lists can be used when search time is not important
- scrollable lists do not have to be used when search time is important
- when search time is important, as many options and levels can be placed on a single menu

Menu titles should

- be meaningful and evident
- provide separate titles for multiple menus or option groups
- use sans serif fonts
- use high contrast against background

- provide visual cues to indicate multiple selection in a consistent location and manner
- visually highlight letters used for implicit designators to be distinct and different from the reminder of the designator name
- be descriptive
- be distinctive
- be compoundable
- provide visibly selected option names in a hierarchy
- provide obvious numbering schemes
- provide obvious and consistent graphic techniques
- provide apparent hierarchy relationship between panels
- allow menu maps that represent the menu structure and be available to the user on demand

10. Navigation should be designed to

- provide cues to help user learn menu structure and navigate
- present menus in shortest time possible
- present menus in 2 seconds or less
- allow navigation from one node of a menu tree to another without returning to initial common node
- allow skipping of intermediate levels of a tree if compatible with task requirements
- provide simple, one step action to return to initial menu
- provide single action to move to next higher level in menu
- provide multiple pathways to access menu levels if logical

11. Option groups should be designed to

- normally limit the number of options to 8 or less
- group options by functions or into logical categories
- arrange options into equally distributed groups
- place options consistently in same order within a group
- list options in conventional (widely used) order with most important options first
- for small options groups, place most frequently used options first
- use alphabetic ordering as last option

Option selection and activation should

- provide alternative methods for option selection
- use separate actions for selecting and activating menu options
- provide bypass intermediate menus
- provide method to combine selection and activation
- provide feedback on selected option
- provide means for selecting and unselecting options
- provide visual cue that menu options has been selected
- allow user to choose options and change options before activation

12. Pointing selection/activation should be designed to

- make the size of the box or button at least twice the size of the selecting symbol if a selection box or radius button is in an unlabeled adjoining area
- provide sufficient separation between selectable areas to minimize unintended activation
- provide feedback upon activation
- provide keying method for selecting and executing options
- provide an "undo" option
- continually display critical options
- display options that are continually or frequently needed in accessible locations not obscuring task data
- present options not frequently needed in pop-up or pull-down panels or in a dedicated area of the screen
- present only options available when unavailable options are not required
- visually highlight unavailable options that can become available
- automatically place cursor at options that have a high probability of selection
- place cursor at the first option in the group if repetition of the option selection is not considered important
- place cursor at the last selected option if the repeating last option previously selected is important
- place cursor at least destructive option if activation of other options can result in destructive actions

Software Checklist

- Consistently locate screen and window titles.
- Consistently locate:
  - page numbers
  - function key labels
  - data and time
  - system name
  - system status
  - window controls
- Display information, functions, and controls necessary for the task.
- Display data in a format immediately usable.
- Display format as part of the data entry fields.
- Display values and ranges acceptable to the system next-to-input field.
- Minimize density of data on the screen display continuous text (long character strings) in upper and lower case.
- Use abbreviations only when necessary and not for system messages.
- Use standard method or symbols to indicate user-entry prompts.
- Use highlighting to call attention to critical information or to indicate item selection.
- Indicate when more information is available.
- Provide method to page or scroll through the material in all necessary directions.
- Use graphs rather than tables and text to show complex, dynamic relationships or changes in data or status.
- Provide screens, windows, and menus with titles that are meaningful to the current task and consistent within an application.
- Provide multi-window systems with appropriate window defaults that users can change easily without programming.
- Clearly indicate the focus or active window at all times.
- Allows users to move, size, open, and close windows easily by using consistent methods throughout the interface.

- Clearly indicate the status of applications running in each window or screen.
- Make all icons discernible from their background and from other icons.
- Make icons easily interpretable.
- Provide an obvious and consistent way to undo critical actions (at least for the last action).
- Allow exiting data entry screens and input windows that don't require data input, except when critical system error or data loss will results.
- Allow free movement between data fields.
- Provide single-action field delete and character by character delete in text and data fields.
- Accept input in any case and display it as input.
- Minimize the use of modes.
- Make same action produce the same results.
- Indicate current mode clearly, and prominently minimize keystrokes and other input device actions.
- Provide automatic computation of data.
- Allocate frequent functions to hard functions keys .
- Function key assignment should be consistent commands, and data entry sequences should match expected sequences for critical tasks.
- Minimize the need for special character input.
- Make commands task-oriented and meaningful.
- Make selectable areas for screen items greater than twice the active area of pointers or cursors.
- Make each type of software control look and act distinctively.
- Action-object syntax should be used.
- Provide multiple methods for user input and navigation.
- Error messages should be brief and informative.
- Error messages should describe specific causes or errors and how to recover.
- Clearly indicate when the system will not accept input.

- Normal system status messages should not interrupt users or require an explicit action.
- On line help and error messages should be relevant to tasks and appropriate to user skill levels.
- Present messages at more than one level of detail to accommodate multiple skill levels.
- Display "help" text without data loss, in a separate window or dedicated area, without hiding the related applications.
- Minimize abbreviations.
- System "help" should contain a list of abbreviations and icons with definitions and results of activation.
- On-line help and error messages should be displayed in plain language at the knowledge level of the user.
- Phrase messages to imply users and not the system is in control.
- Describe procedures in the order that they will be accomplished.
- Provide directions that tell or show users how to continue their task.
- Provide information on the screen when user actions are required.
- Limit initial on-line help to three pages; provide additional information upon request.

Menu Choices:
- Do not present or de-highlight non-available options.
- Put menus in natural or conventional groups, such as
  - important or critical items
  - frequency of use
  - easily learned groups
  - alphabetic order
- Use 4–8 options per menu level.
- Allow access to main menu in one step.
- Make menu organization and terminology reflect user tasks (not software structure).
- Provide an easily accessible method to trace previous actions.

- Design software to automatically check for user errors and allow correction as soon as an error is detected.
- In general, provide feedback to users within 1.0 second.
- Provide feedback for the following actions in the specified time:
  - scrolling and paging: 1.0 sec.
  - function selection: 2.0 sec.
  - interactive requests: 2.0 sec.
  - error messages: 2.0 sec.
  - load or save files: < 10 sec.

Part 5: EU Directive Requirements

The following is an edited summary of conformance requirements from the EU directive *Minimum Safety and Health Requirements for Work with Display-Screen Equipment*. It applies to alphanumeric and graphic display screens, regardless of the display technology. Only those parts of the directive that are applicable to products are included.

Section I: General Provisions

Article 1: Subject. The directive specifies minimum safety and health requirements for work with display-screen equipment.

The directive shall not apply to:

- drivers' cabs or control cabs for vehicles or machinery
- computer systems on board a means of transport
- computer systems not in prolonged use at a workstation
- calculators, cash registers, and any equipment having a small data or measurement display required for direct use of the equipment
- typewriters of traditional design, of the type known as 'typewriter with window'

Section II: Employers' Obligations

Article 3: Analysis of Workstations

1. Employers shall perform an analysis of workstations in order to assess their safety and health conditions, particularly regarding possible risks to eyesight, physical problems, and problems of mental stress.
2. Employers shall take appropriate measures to remedy the risks found, on the basis of the evaluation referred to in the above item 1, considering the additional and/or combined effects of the risks identified.

Article 4: Workstations Put into Service for the First Time. Employers must take appropriate steps to ensure that workstations installed since December 31, 1992 meet the minimum requirements specified in the Annex.

Article 5: Workstations Already Installed. Employers must take appropriate to steps ensure that workstations installed since December 31, 1992 comply with the minimum requirements specified in the Annex and not later than four years after this date.

Article 6: Informing and Training Workers

1. Workers shall receive information on all aspects of safety and health relating to their workstations and in particular information on measures applicable to workstations implemented under Articles 3, 7, and 9.
2. In all cases, workers or their representatives shall be informed of any health and safety measure taken in compliance with this directive.
3. Every worker shall also receive training in use of the workstation before commencing work with display screens and whenever the organization of the workstation is substantially modified.

Article 7: Daily Work Routine. The employer must plan the worker's activities in such a way that daily work with a display screen is periodically interrupted by breaks or changes of activity, reducing the workload at the display screen.

Article 12: Final Provisions

1. Member States shall implement laws, regulations, and administrative provisions necessary to comply with this directive since December 31, 1992 and inform the [European] Commission.
2. Member States shall communicate to the [European] Commission the text of the provisions of national law that they adopt or have adopted on the topics covered by this directive.
3. Member States shall report to the Commission every four years on the practical implementation of the provisions of this directive, indicating the points of view of employers and workers.
4. The Commission shall inform the European Parliament, the Council, the Economic and Social Committee, and the Advisory Committee on Safety, Hygiene, and Health Protection at Work.
5. The Commission shall submit a report on the implementation of this directive at regular internals to the European Parliament, the Council, and the Economic and Social Committee.

Annex: Minimum Requirements

The obligations described in this Annex shall apply to achieve the objectives of this directive and to the extent that the components concerned are present at the workstation, and that the inherent requirements or characteristics of the tasks do not preclude it.

1. Equipment

   a) General comment: The use of equipment must not be a source of risk for workers.

b) Display screen
   - The characters on the screen shall be well-defined and clearly formed, of adequate size and spacing between characters and lines.
   - The images on the screen should be stable, with no flickering or other forms of instability.
   - The brightness and/or the contrast between the characters and the background shall be easily adjustable by the operator and also be easily adjustable to ambient conditions.
   - The screen must swivel and tilt easily and freely to adapt to the needs of the operator.
   - The screen shall be free of reflective glare and reflections liable to cause discomfort to the user.
c) Keyboard
   - The keyboard shall be tiltable and separate from the screen to allow the worker to obtain a comfortable working position avoiding fatigue in the arms or hands.
   - The space in front of the keyboard shall be sufficient to provide support for the hands and arms of the operator.
   - The keyboard shall have a matte surface to avoid reflective glare.
   - The arrangement of the keyboard and the characteristics of the keys shall facilitate the use of the keyboard.
   - The symbols on the keys shall have adequate contrast and be legible from the design working position.
d) Work desk or work surface
   - The work desk or work surface shall have a sufficiently large, low-reflectance surface and allow a flexible arrangement of the screen, keyboard, documents, and related equipment.
   - The document holder shall be stable and adjustable and shall be positioned to minimize the need for uncomfortable head and eye movements.
   - Adequate space shall be provided for workers to obtain a comfortable position.
e) Work chair
   - The work chair shall be stable and allow the operator easy freedom of movement and a comfortable position.

- The seat shall be adjustable in height.
- The seat back shall be adjustable in both height and tilt.
- A footrest shall be made available to the operator.

2. Environment

   a) The workstation shall be dimensioned and designed to provide sufficient space for the user to change position and movements.
   b) Lighting
      - Room lighting and/or spot lighting (work lamps) shall ensure satisfactory lighting conditions and an appropriate contrast between the screen and the background environment, taking into account the type of work and the user's vision requirements.
      - Possible disturbing glare and reflections on the screen or other equipment shall be prevented by coordinating workplace and workstation layout with the positioning and technical characteristics of artificial light sources.
   c) Reflections and glare
      - Workstations shall be so designed that sources of light, such as windows and other openings, transparent or translucent walls, and brightly colored fixtures or walls cause no direct glare and, as far as possible, no reflections on the screen.
      - Windows shall be fitted with a suitable system of adjustable covering to attenuate the daylight that falls on the workstation.
   d) Noise
      Noise emitted by equipment belonging to the workstation(s) shall be evaluated when a workstation is being installed so that it will not distract attention or disturb speech.
   e) Heat
      Workstation equipment shall not produce an amount of heat that could cause discomfort to workers.
   f) Radiation
      All radiation, with the exception of the visible part of the electromagnetic spectrum, shall be reduced to negligible levels for protection of worker safety and health.

g) Humidity
   An adequate level of humidity shall be established and maintained.

3. Operator/computer interface

In designing, selecting, commissioning, and modifying software, and in designing tasks using display screen equipment, the employer shall consider the following principles:

a) Software must be suitable for the task.
b) Software must be easy to use and, where appropriate, adaptable to the operator's level of knowledge or experience; no quantitative or qualitative checking facility may be used without the knowledge of the workers.
c) Systems must provide feedback to workers on their performance.
d) Systems must display information in a format and at a pace that are adapted to operators.
e) The principles of software ergonomics must be applied, in particular to human data processing.

 Part 6: Overview of Legal Requirements of the ADA

1. Employers with fifteen or more employees must be in compliance.
2. ADA prohibits employment discrimination against "qualified individuals with disabilities," an individual with a disability who meets the skill, experience, education, and other job related requirements of a position held or desired, and who, with or without reasonable accommodation, can perform the essential functions of a job.
3. Persons who currently use drugs illegally and persons with behavior disorders are not protected by the ADA.
4. Employers cannot discriminate against people with disabili-

ties in regard to any employment practices, terms, conditions, and privileges of employment.
5. The ADA specifies types of actions that may constitute discrimination.
6. The ADA defines reasonable accommodation as any change in the work environment or in the way things are done that results in equal employment opportunity for individuals with a disability.
7. An employer must make a reasonable accommodation to the known physical or mental limitations of a qualified applicant or employee unless it can show that the accommodation would cause an undue hardship on the operation of its business. ADA defines undue hardship as an action that is excessively costly, extensive, or disruptive or that would fundamentally alter the nature or operation of the business.
8. An employer may require that an individual not pose a "direct threat" to the health or safety of himself or herself or others. A health or safety threat can only be considered if there is "a significant risk of substantial harm."

Part 7: Guidelines for User Interfaces for Individuals with Disabilities

These guidelines are minimum special features that should be provided for individuals with disabilities. Additional information can be found in *Considerations in the Design of Computers and Operating Systems to Increase their Accessibility to Persons with Disabilities.*

Guidelines for Moderately Disabled

❑ Provide an optional (sequential) mode of operation for input devices that require multiple simultaneous activation that is available at any time eliminating the need for simultaneous actions.

❑ Provide a time-internal adjustment, or non-time-dependent response method, so that systems requiring responses in less

than 5 seconds, or a release of a key in less than 1.5 seconds, can be adjusted to allow a longer user response.
- ❏ Provide a method for carrying out mice or other pointing device functions with the keyboard.
- ❏ Allow media insertion and removal with minimal reach and manual dexterity on removable media drives.
- ❏ Make controls (and labels) that are required for periodic system operation accessible and operable with minimum dexterity.
- ❏ Provide a special option (difficult to invoke accidentally) that would delay the acceptance of a keystroke for a preset, adjustable amount of time, and/a keyguard or keyguard mounting provision.

Guidelines for Severely Disabled

- ❏ Provide externally available, industry standard connection point(s) (standard or special port(s)) for adaptive input devices; ensure that the computer accepts input from the adaptive devices the same as it would from other standard input devices such as a keyboard, mouse, or tablet.
- ❏ Ensure that the ability to simulate input-device activities be available to programs running in the background on computers that support background processing or multi-tasking.

Guidelines for the Visually Impaired

For Moderate Impairments

- ❏ Provide a means for
  - ❏ attaching larger (and movable) displays
  - ❏ enlarging images on the display
- ❏ Allow user selection of colors when the color of graphics or text must be distinguished in order to understand information on the display.
- ❏ Provide easily legible lettering on keys and controls.

For Severe Impairments (Total Blindness)

- ❏ If a computer has a standard input system that requires continual visual feedback to operate (e.g., mouse, touchscreen),

provide an alternative means or mode for achieving as many functions as possible. Allow the alternative means or mode to be available at any time, and do not require continual visual feedback.
- ❏ Provide, or make available, a non-visual indication of the state of the toggle keys.
- ❏ Provide keyboards/keypads with tactile discernible key edges (e.g., no flat membrane keyboards without ridges).
- ❏ Provide a distinct tactile marking on the home keys for keyboards and keypads.
- ❏ Provide optional or built-in non-visual key labeling.

Guidelines for Hearing Disabled

- ❏ Present all auditory information that is required for system operation and error detection in a redundant appropriate visual form.
- ❏ Provide user-adjustable and/or reasonably loud volume.
- ❏ Auditory information should also be easily available in a clear form for amplification.

Guideline for Users with Seizure Disorders

- ❏ Avoid refresh or flashing frequencies that are most likely to trigger seizure activity.

Supporting Guidelines

- ❏ Make manuals and other important documentation available in electronic form.
- ❏ Provide built-in speech-output capability or make it available by connecting a speech synthesizer to an output port.
- ❏ Provide windowing environments with the ability to open and maintain special windows that can remain always fully visible.
- ❏ Provide a standard way to connect at least two momentary contact input switches.
- ❏ Provide a means for distinguishing between typed, auto-repeat, and macro-generated "keystrokes" so that they can be

treated differently by the operating systems and application software.
- ❏ Facilitate keyguard mounting.

Product Feature Checklist for Disabled Users

Table 8–23 Specific guidelines for users with disabilities

Topic	Guideline
Reach	❏ Minimum reach distances
Posture	❏ Easily operable without excessive bending or twisting
Button force	❏ Low pressure push buttons (less than 100 grams of pressure)
Button shape	❏ Concave push buttons
	❏ Sliding or edge-operated controls
	❏ Up/down (integrating) control buttons
	❏ Double-action (pop-up, lock-down) push button controls
	❏ Rocker switches (concave)
	❏ Controls operable with minimum dexterity
	❏ Controls which do not require twist or push-and-twist in combination
	❏ Push buttons with tactually discernible edges
	❏ Embossed control indicator symbols
Location	❏ Front edge of the equipment
	❏ Controls which do not require the user to lean around the side or back of the equipment to see or operate the controls
	❏ Control near tactile landmarks (like raised edge of control panel)
Layout	❏ Spatial groupings of controls, especially similar function controls, that provide landmarks
	❏ Maintain standard layout
	❏ Overlays that provide tactile feedback of label differences
	❏ Braille map of controls and brief meanings
	❏ Simple to understand labels
	❏ Simple language on labels
	❏ Consistent labeling
Letter height	❏ Label letters 5 mm in height

Table 8–23 *(Continued)*

Topic	Guideline
Contrast	❑ Light color text on dark background
Color	❑ Red and aqua (turquoise) indicator lights
	❑ Magenta and green indicator lights
	❑ Avoid red and green lighted displays
Color contrast	❑ Highly saturated and high contrast colored lights for displays
Font	❑ Bold fonts for labels
Blinking	❑ Display indicator flashing rate less than 5 HZ or more than 30 Hz
Redundancy	❑ Present auditory information visually
Warning indicators	❑ Warning beeps and tones with visual indicators (flashing/changing color)
Volume	❑ Provide volume adjustment
Association	❑ Provide warning sound associated with moving parts
Sound level	❑ Beeping tones below 750 Hz

Reference

1. The items listed in Tables 8–2 to 8–17 is a condensation of items in ISO 9241, Parts 1–17.

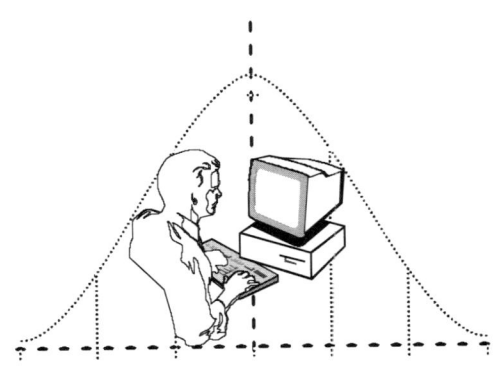

Chapter 9
Usability Testing

Overview

Certain ergonomic standards, such as ISO 9241, allow compliance of a product to be demonstrated if it can be shown that user performance, comfort, and effort are not significantly worse than that of a product that meets ergonomic standards, or that meets a de facto standard currently on the market. This chapter briefly describes examples of hardware and software user interface testing that can be used for this purpose. The test procedures are based on ergonomic practices and testing methods and measures in the ISO 9241 VDT ergonomic standard.[1]

The first part of the chapter describes general information applicable to all the tests; the following sections provide information specific to a particular test. Icons indicate information specific to displays, keyboards, non-keyboard input devices, or software (see Table 9–1).

Background 239

Table 9–1 Testing reference icons

Information for testing of:	Indicated by:
Keyboards	⌨
Non-keyboard input devices	🖱
Displays	💻
Software	💾

Background

Design Test Measures

Usability testing is a necessary component of ergonomic practice and standards compliance analysis because design specifications are not sufficient for a total assessment of the product's impact on user performance, effort and comfort.

In order to evaluate effectively the ergonomic quality of a product, it is necessary to measure all these variables simultaneously. None of them alone is a sufficient indicator of the ergonomic quality. For example, products or components that may conform to design specifications and result in acceptable user performance may result in discomfort and stress in an effort to meet or maintain performance goals. As biomechanical effort increases, discomfort eventually results, and performance will ultimately be affected. When biomechanical effort is excessive and continuous, long-term disorders can occur. For example, most computer users can key efficiently using a standard keyboard or accurately manipulate images on a screen with a mouse. However, ergonomic and medical studies show that recommended biomechanical limits can be exceeded (see Figure 9–1) when using these devices—even though they may meet traditional design specifications. In addition, the postures that result from using these input de-

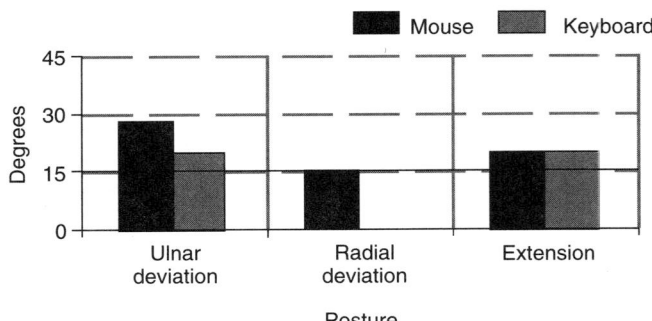

Figure 9–1 Average keyboard and mouse hand angles. Neutral posture is 0°. Continuous or frequent deviations greater than 15 degrees (indicated by solid marker line) result in biomechanical stress.

vices can increase the risk of cumulative trauma disorders (see Chapter 6).

Biomechanical and Physiological Test Measures

However, although minimal biomechanical effort should be a design goal and is necessary for continuous performance, it may not result in optimal performance, as shown in the following examples:

 Research has shown that the force required to depress some keys may be so low in some cases that users will key with more force than is required and as a result may make more errors. Wrist rests may reduce arm load during keying for some users. However, they can impede hand and arm motions during high speed keying, reducing performance, and are thus not usually preferred by high speed keyists. In addition, wrist rests that are not padded can cause wrist injuries.

 A display may meet an ergonomic design specification, such as a refresh rate of 60 Hz, but still result in flicker that is perceptible to a large number of users. If sufficiently disruptive, the flicker will eventually reduce visual performance and cognitive processing.

Background

 A software dialogue can meet general semantic specifications, but if the semantics are not geared to the skill level of the user, it will result in cognitive load, frustration, stress, and eventually reduction in performance.

Performance and Comfort Test Measures

Using performance alone as an indicator of ergonomic quality is also an insufficient measure (see Figure 9–2), because excessive effort and discomfort can occur even though short term studies may show acceptable performance. Usability studies have shown that performance is not highly correlated with comfort ratings. The subject's impressions of performance and comfort are often not accurately reported and perception of comfort is often unreliable and inconsistent. Most usability tests are not of sufficient duration for users to perceive and accurately report performance and comfort effects. Even if discomfort is reported, it cannot be relied on as an accurate effort indicator because users are often not aware of the state of their physiological processes, such as nerve compression, reduction in blood circulation, and oxygen deprivation. However, effort can be accurately measured with biomechanical instrumentation.

Multiple Test Measures

Thus, product design and selection should be based on the simultaneous measurement of several variables. This approach offers

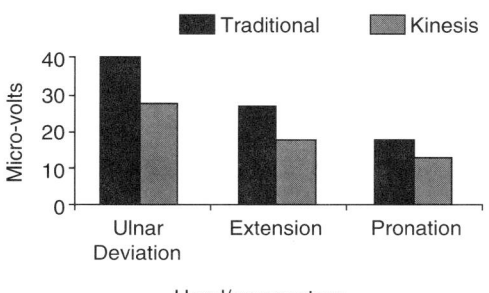

Figure 9–2 Test data showed that the Kinesis keyboard resulted in lower muscle load although performance was slightly better on the Traditional keyboard

an opportunity to observe problems that are not evident with single measures and during the short duration typical in usability testing. Simultaneous testing of multiple measures also minimizes test time and cost and allows correlation of performance, comfort, and effort.

Each measure of usability is assessed in a different manner:

- *Performance* is measured by speed and accuracy.
- *Biomechanical effort* is assessed by measuring muscle load, muscle fatigue, maximum voluntary contraction, and postural angles.
- *Physiological functioning* is measured by monitoring specific functions like blink rate, visual accommodation, heart rate, galvanic skin response, and blood pressure.
- *Comfort* is assessed with rating scales, comments, and responses to questionnaires.

Performance, biomechanical load, and physiological functioning can all be quantified and empirically measured. Selection and use of measures depend on the product being tested and the test objectives. Table 9–2 shows which measures should be used to test different products.

Table 9–2 Product usability test measures

	Measure				
Product	Performance	Posture	Muscle Effort	Physiological Response	Comfort
Keyboard	✓	✓	✓		✓
Non-keyboard input device	✓	✓	✓		✓
Display	✓	✓		✓	✓
Software	✓			✓	
Furniture		✓	✓	✓	✓
Environment				✓	✓

 ## General Test Conditions and Procedures

The following is an overview of general test conditions and procedures for computer hardware and software.

 Test Administrator

Testing should be conducted by a professional who is trained and skilled in usability test methodologies, psychometrics, and statistical analysis. The tester should have graduate training in experimental psychology and human factors engineering/ergonomics. The tester should have additional training for tests of the following products:

 For input device tests—a background in biomechanics, physiology and the use and calibration of instruments to measure these functions.

 For display tests—a background in visual psychophysics.

 For software tests—a background in cognition and linguistics.

Knowledge of these additional areas is important to determine appropriate test tasks, functions to be monitored, data to be collected, and interpretation and application of test results.

The test administrator should be available to the subjects throughout the test to answer any questions that arise before and during the test.

 Test Subjects

Subject Characteristics. Test subjects should be representative of the anticipated user population(s). User variables such as gender, age, visual ability, handedness, hand size, manual dexterity, product experience, and skill level should be identified and used as subject selection criteria.

 For a keyboard test, subjects might include users with different manual abilities and problems (with and without RSI wrist disorders).

 For a non-keyboard input device test, subjects could have different hand dominance (left or right) and different levels of familiarization and skills.

 For a display test, subjects could have different visual abilities (color vision deficiencies and normal color vision).

Number of Test Subjects. A sufficient number of subjects should be tested to establish a statistical confidence level of at least .95. Although this usually requires a test population of not less than twenty-five subjects, the exact number depends on a variety of factors such as subject grouping, number of trials per subject, variability of subject's responses and desired confidence level of the test.

Test Forms

Relevant authorization forms (for such matters as for test participation, video releases, and nondisclosure agreements) should be completed by all subjects. Confidentiality of personal information, questionnaires, performance and test scores must be assured. Information which reveals test subject identity should not be released by the test agency/administrator.

General Test Conditions and Procedures   **245**

 Test Room

Testing should occur in a dedicated facility such as a usability laboratory and be free of distractions. The advantages of a usability laboratory are that it provides a separate observation room and a means of communication with the test subject without the observer having to enter the test room.

A usability lab also can provide an environmentally controlled test environment. The test room should meet ISO 9241 environmental specifications (see Table 9–3), or replicate the conditions in which the product is expected to be used. For example, in testing products for CAD use, the light level might be set to 200 lux, which is common in CAD work environments. If the test room conditions vary from ISO specifications, the reason for this should be described in the test report.

 Test Station

The test station (that is, the table, chair, document holder, footrest, and auxiliary input device support) should meet ISO 9241 specifications, which include the following:

Table 9–3 ISO 9241 test room requirements

Room Condition	Requirement
Ambient noise	< 55 dB(A)
Temperature	19° to 26° C
Humidity	40% to 60%
Air velocity	< 0.15m/s
Illumination	500 lux

- computer device support surface with adequate space for the display and input device use
- separately adjustable input device and display surface heights
- display tilt mechanism
- chair with an adjustable seat height, back angle, and a stable base

The test station should allow products being tested to be located in their intended operating locations and positions.

Test Equipment

 Product Identification

Vendor identification marks, such as corporate logos and names, should be concealed on the products being tested. Test products should have identifiers that are anonymous (for example, hardware devices can be labeled "A" and "B"; software user interfaces can have anonymous identifiers in their title bars).

Product Setup and Reporting

 Test displays (and reference displays, if used) should be put into operation long enough in advance of testing (and any training sessions) that their functions will be stabilized. The same screen polarity should be used to display user inputs and actions across test conditions. Anti-glare and anti-reflection devices that are intended to be marketed with the test display should be used throughout the test. The chromaticity of the ambient test room illumination should be identified and described in the test report.

 If the test keyboard is typically sold with a specific display, that display should be used with all the keyboards tested. A description of keyboard features like size, slope, layout, and unique design features should be included in the test report.

The same display should be used for displaying the results of input device activations. The type and material of the input device support surface should be described. A description of the size and unique design features of the devices should be included in the test report.

Reference products should meet ergonomic standards and provide the functions necessary for subjects to accomplish the required tasks.

Monitoring Equipment

 Computer System

A computer system is necessary to display test information and monitor users' interactions with products. The system should be capable of:

- displaying input device activations (such as characters keyed, cursor movements, and image manipulations) and visual images on the test subject's display
- automatically capturing and storing interactions in sequential order
- timing subject's actions in milliseconds
- displaying subject actions on an observation display
- statistically analyzing interactions, including biomechanical and physiological measures and posture angles
- identifing and classifing errors

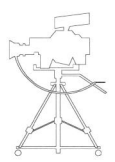 Video System

A video recording system is also necessary and should be capable of

- videotaping and displaying subjects' interactions (such as keying, mousing, posture and movements) on an observation display
- time stamping in milliseconds
- character generation for subject and session labeling
- instant replay
- sufficient focus or image conversion to allow good visibility of screen images, subject actions, and postures
- recording from at least two different views
- simultaneously display of different camera views on the observation monitor
- fading, positioning, and zooming
- recording on either VHS or Super VHS videotape
  Note: Recording of screen images on VHS videotape may not result in sufficient quality be able to read the characters of the screen.

 Nonrelated Software

System software unrelated to the test should not be allowed to interfere with the test and measurements. For example, if the test computer system is connected to a network, unrelated messages, such as notification of the system status, should be disabled.

Test Procedures

Subject Treatment

It is essential that rules governing the ethical conduct of human testing be followed. For example, subjects should begin the test in a relaxed state and should not be exposed to unnecessary or excessive stress.

The same test instructions should be given to each subject. The instructions should inform subjects to work as quickly and accurately as possible and to leave errors uncorrected, unless error correction is part of the test. Subjects should be allowed to become adapted to the test room for at least fifteen minutes prior to the beginning of the test.

All subjects should be tested using all products. Assignment of subjects to products should be counterbalanced to eliminate order effects (see Table 9–4).

Pretesting

 The keying proficiency of the test subjects should be evaluated prior to the start of the test. Subjects' keying proficiency should represent the level of the intended users. The minimum recommended keying proficiency is typically 45 words per minute with no errors. A standard typing test should be used whether or not subjects meet this criterion.

 Display test subjects should be given a vision test to insure they can see the screen images sufficiently well to participate effectively in the test. Standard tests to measure visual acuity and color vision should be used.

 Software test subjects should demonstrate their ability to use the operating system(s) and input devices to be used in the test.

Subjects should demonstrate their familiarity with the product(s) to be used in the test and be given sufficient time to use them until it becomes obvious that they understand their proper operation and their performance shows no significant improvement.

Figure 9–4 Example of balanced subject assignment

Subject	Product use sequence	
1	A	B
2	B	A
3	A	B
4	B	A
5	A	B

For novel keyboard designs, this usually requires four to eight hours.

 For novel non-keyboard input devices, familiarization time depends on the complexity of the device and difference from other devices they have used. It could require 10 minutes to several hours.

 Learning software can require up to several weeks depending on its complexity and the level of knowledge required for the test tasks.

If one of the test goals is to determine how long it takes to become efficient in using the product, this data can be collected during the training session.

The types of tasks for which a product was designed should be used in selecting the test tasks.

Video recording and monitoring of any biomechanical/physiological functions should be obtained while subjects are performing the test tasks.

Test sessions should be of sufficient length to obtain statistically valid data. Ideally, the test should not exceed four hours per subject per day, including breaks. The number and length of breaks depend on the test. Subjects should have at least a 5 minute break every hour and a 10 minute break every two hours.

 **Keyboard Test**

Keying Input

Subjects should key text or alphanumeric characters presented in black ink on white paper. Character type font should be a style that is easily legible and familiar to the intended user population (see Table 9–5). Fancy, italicized, or very small fonts should not be used.

Keyboard Test

Table 9–5 Examples of acceptable and unacceptable input characters

Character feature	Acceptable	Unacceptable
Fonts	Times New Roman	*Ashley Script*
	Courier	Fenice
	Century schoolbook	**Boecklin**
Style	Regular	*Italics*
		Bold
Size	10 pitch	≤8 pitch
	or 12 pitch	≥14 pitch
Contrast	≥ 10:1	≥ 10:1
Color	Black	Color

There should be a sufficient number of input sheets so that test subjects do not repeat entry of the characters throughout the test.

The input characters to be keyed can be presented in continuous text or in random alphabetic and numeric characters. Continuous text should be:

- upper and lower case
- the same for each keyboard tested
- in the usual language of the intended user population
- at the reading ability of the intended user population (eighth grade reading level is recommended)
- neutral in content and not too technical nor scientific
- free from spelling and grammatical errors and correctly punctuated
- void of any numeric characters or special characters such as a pound sign, asterisk, ampersand or special features such as italics, boldface, or underline
- double-spaced
- matched to the images on the display as closely as possible (that is, the displayed text or data should match the data sheets in terms of font spacing, line width, line justification)

Table 9–6 Examples of input data

Alphabetic	Numeric	Alphanumeric
SOENFIL	2017947	FD5U8G1
OAPICAI	9329450	89I4F4H
TOZNBHT	1623437	X3H89ID
MTODSRI	1361489	5F9EKIJ
EIFRESG	2756490	J7K4F3F

- void of indentations
- different for each test trial but in the same sequence for all test subjects

Random characters should be in upper case; they can be all alphabetic, numeric, or a combination (see Table 9–6). Random character input sheets should consist of five vertical groups of five lines of characters with seven characters in each line.

Keying Tasks

Keying should be limited to the alphanumeric or numeric portion of the keyboard. The test excludes interaction with, and thus evaluation of, function keys and other aspects of keyboards that are dependent on, and controlled by, custom software.

Keying should occur for six 20-minute sessions with a 5-minute break between each session except for a 15-minute break between the third and fourth sessions (see Table 9–7).

Subjects should be allowed at least 5 minutes of practice keying on each keyboard. When practice sessions are not possible, the first two sessions should be considered practice. Thus the data from these two sessions are not normally included in the performance analysis, unless learning time is to be measured and reported.

Keying Performance Analysis

The total number of words or characters keyed for each 20-minute session and the number of incorrect words and incorrect characters should be calculated. Keying throughput should be cal-

Table 9–7 Test session schedule

Session	Activity	Duration (in min.)
1	keying	20
	break	5
2	keying	20
	break	5
3	keying	20
	break	15
4	keying	20
	break	5
5	keying	20
	break	5
6	keying	20

culated by dividing the number correct in each session by the number of minutes in the session.

 ## Non-keyboard Input Device Tests

Test Task Selection

Most non-keyboard input devices are used for a variety of tasks like pointing, selecting, dragging, tracing, and freehand drawing. Six tests have been designed to evaluate these tasks. The tests used should be determined by the tasks expected of the intended user population (see Table 9–8).

All subjects should have the same test conditions. All the tests should be conducted with a range of different difficulties.

Pointing Tests

Pointing tests include a one-direction tapping test and a multi-directional tapping test.

1. One-Direction Tapping Test. The one-direction tapping test requires subjects to move a cursor along one axis from one rectangle

Table 9–8 Selection criteria for input test(s)

For generic tasks like:	And applications like:	Use:
Pointing and moving cursor along one axis	• horizontal / vertical rubber-banding • inserting a cursor at point along a character string • selecting information in columns	One-direction tapping test
Pointing in different directions	• repositioning cursor at different areas • selecting cell in spread sheet • selecting icons along horizontal and vertical tool bars	Multi-directional tapping test
Clicking and dragging to specific locations	• clicking and dragging the cursor down a pull-down menu • inserting, clicking and dragging cursor along a string of text to highlight it • dragging an object from one window to another	Dragging test
Clicking and dragging objects to specific locations or duplication shapes	• tracing an image (like a PCB layout) on a tablet • duplicating lines or shapes area filling of objects	Tracing test
Hand drawn images	• graphics creation • free hand entry and character recognition operations • character recognition hand-held pen based-systems.	Freehand input test
Keying and input device use	• text editing graphic design • numeric data entry in a spreadsheet	Grasp and park test

to another and activate the input device button (click) each time the cursor arrives at the target rectangle (see Figure 9–3). The size of the rectangles and distance between the rectangles should be varied between test trials. There should be 25 trials in each test session. Each session should begin when the user first moves the cursor into the left rectangle and activates the input device button.

Feedback should be provided so that a subject can distinguish between a "hit" and a "miss". If a miss occurs, the subject should repeat the task to obtain a correct hit before continuing.

2. Multi-Directional Tapping Test. The multi-directional tapping test requires subjects to move a cursor across a circle to sequential target squares (see Figure 9–4). The size of the circle and thus the distance between the target squares should be varied between trials. The target squares should be equally spaced around the circumference of the circle and should be arranged so that cursor movements are nearly equal to the diameter of the circle. The target square to which the cursor should be moved should be highlighted (with color, area fill, or brightness).

Each trial should start when the subject points to the topmost square (position "0") and end when the sequence around the circle is completed and the cursor is again positioned at the "0" square.

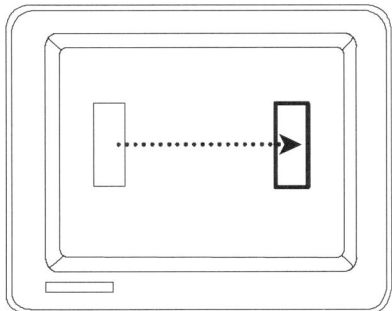

Figure 9–3 One direction tapping task. Movement of cursor (arrow) to target rectangle (on the right)

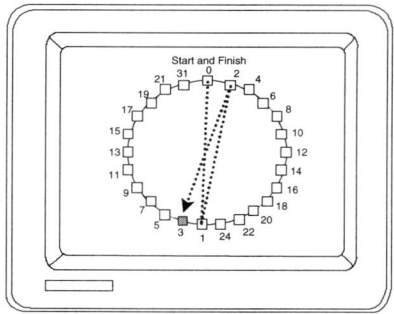

Figure 9–4 Multidirectional tapping task. Movement of cursor to #3 target

3. Dragging Test. The dragging test requires the subject to move an object (like a circle) between a track of two parallel lines without touching them (see Figure 9–5). The width of the track and size of the object should be varied between trials. If the object touches a boundary line, the system should record the action as an error and the subject should be required to start again. The time taken for the subject to move the object successfully from one end of the track to the other should be recorded.

4. Tracing Test. The tracing test requires the subject to move an object within a circular track, without touching the boundary

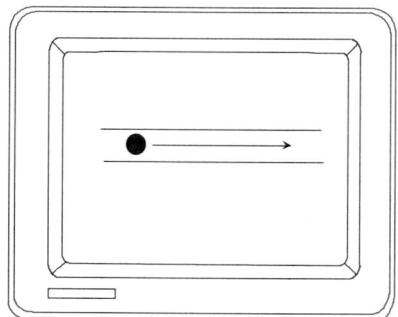

Figure 9–5 Dragging task. Movement of target through the track

lines (see Figure 9–6). The size of the object, width of the track, and circumference of the circle should be varied between trials. If the object touches a boundary line, the system should record the action as an error and the subject should be required to start again. The time taken for the subject to move the object around the track successfully should be recorded.

5. **Free-hand Input Test.** The free-hand input test requires the subject to write or draw a legible symbol in each of a horizontal string of boxes (see Figure 9–7) as rapidly as possible. The size of the boxes and the distance between them should be varied between trials. The time taken to complete the task should be recorded and compared between input devices. The same shapes should be used with each device tested.

This test is intended to compare drawn characters or symbols created with an input device with those drawn with traditional input devices such as a pen or pencil on paper.

6. **Grasp and Park (Homing) Test.** The grasp and park (homing) test requires the subject to use a non-keyboard input device to move the cursor to a specific location on the screen and then to use a key on the keyboard to secure ("click") the cursor into place. The time difference between this task and the equivalent pointing task is the "grasp and park" time.

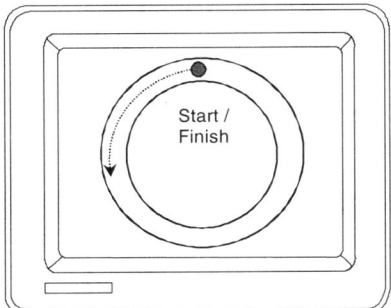

Figure 9–6 Tracing task. Movement of the target around the track

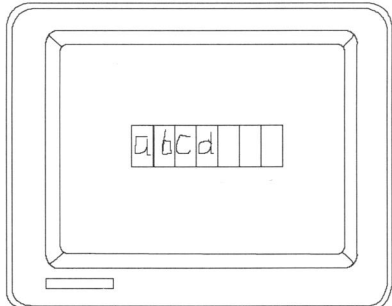

Figure 9–7 Freehand Symbol Entry task. Entry of hand-drawn characters into boxes

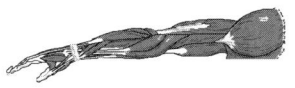 **Biomechanical Assessment**

Type of Measures

There are several ways to measure biomechanical effort during input device use. The measurement method chosen should not interfere with the test tasks nor be intrusive. The non-intrusive methods most frequently used in biomechanical assessments are measures of

1. posture deviations from neutral
2. muscle effort (load or muscle fatigue)
3. strength (maximum voluntary contractions, or MVCs)

There are positive and negative aspects to each method:

1. Posture deviations from neutral must be carefully measured and are best assessed by automated equipment such as electronic goniometers. Manual methods of measuring postures are inexpensive but less accurate than automated systems.

    Posture measures provide only a partial representation of biomechanical effort. Subjects can be displaying neutral postures but be exerting excessive muscle loads while holding an input device or depressing a key or button.

2. Muscle effort measurement requires specially trained administrators and sophisticated instruments which require careful calibration. Both postural deviation and exerted muscle force can be determined from muscle load analysis. However, angular measures of deviation (in degrees) cannot be determined.

   The technique used most often to measure muscle effort is surface electromyographic (EMG) recording. EMG recording is empirically related to the force of exertion for a given posture and has been validated in three decades of testing and use. Like automatic measures of postures, surface EMG recording can be obtained while test subjects are performing their tasks.

3. Strength measures require knowledge and use of strain gauges. In addition, the measurements interrupt testing and are a less valid measure of biomechanical load because they depend on a subject's understanding and motivation to exert maximum voluntary contractions (MVCs). Strength measures may not be indicative of the effects of using input devices because subjects may be able to exert very high strength even though their muscles are stressed and fatiguing. On the other hand, their muscles may not be stressed, but subjects may be so tired (or bored) from the test that they may not exert an MVC. Thus strength is often not a reliable measure of muscle load or fatigue.

Although each of these measures has disadvantages, they should all be used because of their complementary effect and different indications. It may be desirable to compare biomechanical measures between two devices of the same type but of different design (such as two different sizes or shapes of mice) or two different types of devices (such as a mouse and a trackball). Or, it may be desirable to measure biomechanical effort for a specific input device that is used for a variety of tasks. The most valid biomechanical load comparisons are measures from identical muscle groups and the same postures across devices.

Typically postures and muscle effort from the first and last test sessions are excluded from data analysis because of the data contamination from starting and stopping activities.

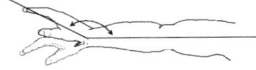 Posture Measurement

Appropriate postures to measure during testing of input device use include hand extension, flexion, and deviation (ulnar and radial), arm pronation, suppination and abduction, and finger separation (see Glossary). Postures should be monitored throughout the test. Measures should be made at least every two minutes.

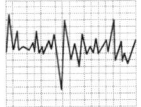 Muscle Effort Measurement

Examples of muscles typically monitored during input device testing are shown in Table 9–9.

Muscle effort measurement involves special procedures, including

- careful set up and periodic calibration of monitoring equipment
- identification and location of muscle groups to be monitored
- preparation of the surface of the test subject's skin for electrode placement
- placement of electrodes and ground on test subject

If muscle load is to be determined, an analysis of *amplitude* of muscle contraction is necessary; if muscle fatigue assessment is desired, the *frequency* of muscle contraction should be analyzed. These analyses require a knowledge of advanced mathematics and statistics.

Table 9–9 Limb actions and controlling muscles

To measure effort during:	Measure activity of:
extension	*extensor communis digitorum*
ulnar deviation	*extensor carpi ulnaris*
arm pronation	*pronator radii teres*
arm abduction	*deltoid*

☒ Comfort Assessment

Comfort should be determined by requiring subjects to fill out rating scales. Subjects should complete an independent questionnaire after using each device and a comparative questionnaire after using all the devices. Subjects should place a check under the number than best indicates their opinion of the characteristic of the particular input device. Low numbers indicate negative ratings; high numbers positive ratings. Either a five-point or seven-point rating scale should be used. Appropriate statistics (like nonparametric statistics) should be used to analyze the resulting data.

The *independent questionnaire* (see Table 9–10) should be used to obtain an assessment of each device independently and should be given to each subject immediately after the input device is used.

The *comparative questionnaire* (see Table 9–11) should be used to obtain an assessment of how the devices compare to each other and should be given to the subjects after they have used all the input devices.

 **Display Usability Test**

Test Description

The purpose of the display test is to evaluate recognition and discrimination of alphanumeric characters and their colors on a display. Since the test focus is on visual perception rather than on cognitive performance, there are no tasks requiring subjects to remember colors or to interpret their meaning. Cognitive performance tasks are included in software tests (as in Parts 11–17 of ISO 9241; also see next section: Software User Interface Tests).

Display tests consist of the following two tasks:

Table 9–10 Independent Rating Scale

	Least Positive						Most Positive
	1	2	3	4	5	6	7
Force require to depress keys							
Smoothness of operation							
General effort to use the device							
Soreness or fatigue in wrists							
Soreness or fatigue in fingers							
Soreness of fatigue in arms							
Fatigue in shoulders							
Posture during input device use							
Impression of accuracy							
Overall device operation							

1. character and color recognition
2. color matching[1]

Test Stimuli

A matrix of random alphanumeric characters should be presented on the display screen. The matrix should consist of three lines of characters with five characters in each line (see Figure 9–8) in all colors in the default color set. All the center characters should be presented in each font and color of the default set against the default background for the test display.

---

[1]The color–matching test can also provide a user performance test of color uniformity.

Table 9–11 Comparative Rating Scale

	First input device: A ☐ B ☐					Second input device: A ☐ B ☐		
	Least Positive				Most Positive	Worse	Same	Better
	1	2	3	4	5	−1	0	+1
1. Touch/feel								
2. Smoothness								
3. Effort required								
4. Comfort								
5. Fatigue								
6. Posture								
7. Awkwardness								

Pretest

In order to ensure that all subjects are equally familiar with the characters and color names to be included in the test, subjects should be given a pretest until they can correctly name all the characters and colors. The pretest should consist of displaying the characters and a cursor-size block of each color of the default set (e.g., red, green, blue, yellow, orange, cyan, and white/black) against a black background, and then a white background, until all the characters and color blocks are correctly named. If there are any errors the entire set should be presented again, with the order of characters and colors presented randomly.

Test subjects should name each character and color. Feedback should be provided to the subjects indicating a correct or incorrect response. The system should record all responses. When all the responses are correct for all characters and colors, the subject is ready to begin the tests. Subjects should repeat the training until the test criterion is met.

Color and Character Recognition Test. The screen character matrices should be displayed one at a time in one of five locations on the screen (center, upper left, upper right, lower left and lower right) for each trial. Just prior to the presentation of each matrix,

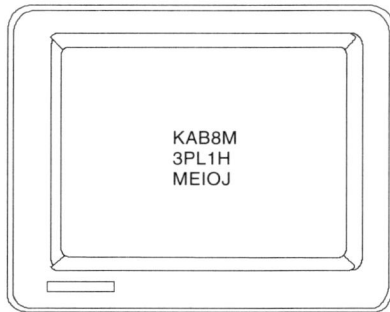

Figure 9-8 Example of test stimuli matrix

a gray block cursor should be displayed briefly (i.e., for 250 ms) to direct the subjects' attention to the portion of the screen where the next matrix will be displayed.

Subjects should name each character and color in the center position of the middle line of the matrix within the 30 seconds at each location. If there is no response from the subject, the trial should be recorded as an error, and the next trial can begin.

The characters, colors and order of the screen positions should be randomized. Incorrect identification or lack of identification of a character or color and corrections of errors should be recorded as errors. No feedback on performance should be given to the subject.

Color-Matching Test. Matrices should be displayed simultaneously at two locations selected randomly from the five test locations (see Figure 9-9). All characters and colors in the default sets should be presented in random order during the test trials.

Subjects should report whether the color of the center character in the two matrices is the same or different. Test matrices should be displayed for a maximum of 30 seconds; if there is no response, the trial should be recorded as an error, and the next trial can begin. Incorrect identification or lack of identification of a color and corrections of errors should be recorded as errors. No feedback on performance should be given to the subject.

This test can also be used to evaluate perceived color uniformity. The uniformity error rate is calculated from subject errors.

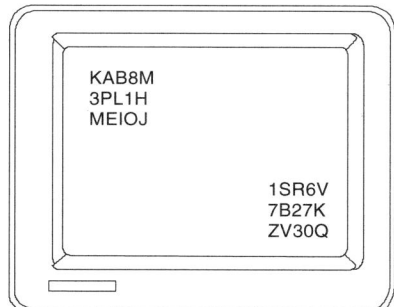

Figure 9–9   Example of locations for color-matching test

Colors are considered uniform if there are no errors in color matching.

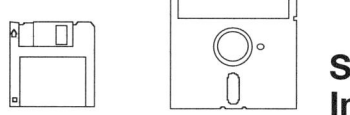 ## Software User Interface Tests

There are several methods for assessing software usability. They include

- system descriptions
- documented evidence
- observations
- analytical evaluations
- empirical evaluations

The method used depends on the interface or feature being evaluated, and testing resources, number of subjects, and number of inspectors available. Since an interface usually includes features that can be quantitatively and qualitatively evaluated and objectively and subjectively measured, all these methods might be useful for evaluating one interface. The ISO 9241 test protocols include all these measures. As with other measures, they each have positive and negative features.

System description is information on system properties and features. It is useful for determining the applicability of the interface to anticipated tasks and for deciding appropriate tests. The description can include user guides, manuals, and on-line helps. This method requires access to information about the computer system, users, and tasks.

Documented evidence is information on the characteristics of anticipated users, tasks, and task sequencing, and test data on other systems which may be useful in assessing the candidacy of a feature for testing.

Observations are inspections of a software product to determine which features are included in the evaluation. They are thus an objective measure but require careful attention and a knowledge of what content should be assessed.

Analytical evaluations are judgments of the quality of interface features, which ones should be evaluated and by which test method. This method is useful during development as well as during product testing. However, it requires inspectors who have sufficient training and skill to make appropriate and accurate judgments about the quality and application of interface features and relative tests. This method is subjective and qualitative.

Empirical evaluations are test procedures assessing user performance in terms of learning time, task accomplishment time, and errors. This method requires controlled testing conditions like those described earlier in this chapter. It requires specially trained test administrators, a test environment, and computer monitoring equipment. This method is most appropriate for evaluating prototypes and final or released products. It is objective and quantitative and, although the most time consuming, is the most accurate and reliable method for assessing the impact of interface features on user performance. Examples of the application of these methods to the analysis of different user interface features is shown in Table 9–12.

Summary

Usability testing demonstrates the ergonomic quality of a product. Since the design of a product should produce the best er-

Summary

Table 9–12 Examples of software test methods to evaluate a usability feature

To evaluate:	System Description	Documented Evidence	Observations	Analytical Evaluation	Empirical Evaluation
Placing of important options first	✓	✓	✓	✓	
Minimization of keystrokes					✓
Cursor response time					✓
Ability to correct before undo			✓		
Emotionally neutral commands		✓			✓
Screen density	✓	✓		✓	
Coding discrimination					✓
Default fields	✓	✓	✓	✓	
Echoing of keying			✓		✓

gonomic features, the tests by which a product is evaluated should be of the highest possible quality. To achieve this goal requires adequately trained test technicians experienced and skilled in planning and conducting tests and appropriately applying the results of these tests.

Unfortunately, usability testing is more time consuming, complicated and costly than measuring the presence or absence of a design feature or measuring its physical characteristics. The advantage of usability testing is that it can provide a more accurate assessment of usability quality than testing product conformance to a quantitative design specification.

This chapter has presented the "tip-of-the-iceberg" of usability testing. Usability testing was included in this book to provide examples of test methodologies, imply its importance and describe some of the testing methods being considered for inclusion in ISO 9241. Inclusion of usability testing in standards is a relatively new concept and has much progress to make until comfort with these evaluation methods is established. Many product manufacturers express concern about the effects of usability testing on product development, release schedules, and budgets. However, if a product's development cycle is appropriately planned, it should include ergonomic analysis and usability testing which should shorten the product life cycle and reduce long term product cost.

Reference

1. The test descriptions are editorials of the test descriptions in ISO 9241, Parts 3, 4, 7.

Chapter 10

The Impact and Future of Ergonomic Standards

Overview

In the last twenty years, concern over the impact of computers on health and safety has led to national and international ergonomic standards, legislation, and litigation. This chapter will discuss the impact ergonomics has had, and probably will have, on product standards, design, development, and marketing. It also includes suggested strategies for product developers to employ in order to maintain a competitive edge.

The Ergonomic Explosion

Ergonomic requirements specify design features that impact the quality of user interaction with products and environments. In order for this interaction to be successful, it must be efficient, easy, comfortable, and safe.

The last ten years have witnessed a plethora of concerns regarding the impact of computer products and environments on worker safety, health, and comfort. This has resulted in the following events:

- Almost all major high technology market areas have ergonomic specifications for computers and office environments.
- Multinational regulations exist in the largest economic organization (the European Union) in the world for the design and use of computer equipment, furniture, and associated work environments.
- Over 200 bills have been introduced into U.S. state legislatures regarding computer product design and use and office environments.
- The United States is in the process of creating federal regulations on repetitive work, which includes work with computer input devices.
- Hundreds of millions of dollars have been lost by computer manufacturers whose products have not met ergonomic requirements.
- Litigation regarding product ergonomics is rapidly increasing, particularly for computer input devices. (In the U.S., worker compensation cases for RSIs have recently risen more than 55%. Thousands of RSI lawsuits have been filed in the U.S.)
- Ergonomic features are becoming a primary competitive strategy and marketing tool.

Although most of these events have occurred in the last decade, this trend is expected to continue for an indeterminate time. Businesses with products that do not comply with ergonomic regulations and standards are becoming less competitive in the marketplace. Failure to address ergonomics and ergonomic standards will increasingly limit the marketability of products. In addition, product purchasers and users are becoming more sophisticated about ergonomics. They are increasingly requesting

- usability test data that validates design features
- data sheets of a product's ergonomic specifications
- proof of compliance to ergonomic standards

In addition, employers and user organizations that want to optimize workplace effectiveness, health, safety, and comfort realize that they can meet these goals if their products conform to ergonomic standards. These groups are becoming aware that products that fail to comply to ergonomic standards ultimately reduce worker effectiveness. In several countries in Europe, codetermination gives employees the right to not work with products they deem unsafe or unhealthy or that do not meet ergonomic standards. Products that do not conform to the EU *Display Screen Directive* (see Chapter 6) will not be purchased by EU employers.

Corporate Strategies

The initial reaction of industry to ergonomic standards was reactive and defensive (see Chapter 2). Industry is slowly learning that it is economically and ethically advantageous to become more proactive in addressing ergonomic product issues and problems. The importance of usability features was made evident to the computer industry in a survey of 6,000 customers, which showed that ease-of-use was one of their top three purchase criteria. The serious attention industry is paying to data like this is demonstrated by increased company investment in improving user interface design and usability testing. For example, the largest software company in the U.S. reportedly spent several million dollars on the development and testing of an ergonomic mouse.

Although other computer companies are beginning to follow this investment trend by yearly budgeting for ergonomic functions, only a few corporations have integrated ergonomics into their product development process. In addition, only major computer corporations generally employ human factors engineers/ergonomists. This practice is even less frequent in other industries such as those that manufacture consumer products and medical and scientific instruments. However, hiring human factors engineers is slowly spreading into these industries because companies that have made a commitment to high quality ergonomic design are gaining global reputations for easy-to-use products and increasing their market share. These companies staff human factors engineers, retain the consulting and testing services of ergonomic institutes, and participate in the development of national and international ergonomic standards. Some prominent examples include Apple Computer, Microsoft, Xerox, NEC, Siemens, Philips, Nokia, ICL, and Ericksson.

Product Strategies

There are a number of strategies that can more competitively position products in a market that increasingly demands ergonomic features. These practices can help companies remain competitive:

- produce products that meet ergonomic principles and standards
- be more responsive to ergonomic issues and concerns of office workers
- demonstrate due diligence regarding appropriate product and work environment design
- comply with standards and regulations

On the following pages are recommended strategies for the ergonomic design and development of several computer-related products.

  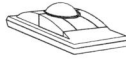 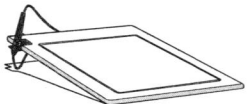

Input Devices

Designers of input devices should increase their awareness of ergonomic and medical research findings and the implications of these findings for product design. Computer manual input devices need to be redesigned to reduce biomechanical load during use. Specifically, input devices should be designed so that

- *their shape fits the hand(s) during use in a neutral posture*
- *drag lock is a standard feature so that dragging screen images is possible without the continuous depression of a button*
- *button/key depression forces are compatible with the findings of recent ergonomic research*

The major risk factors to disorders of the hand, arm, and fingers are

- highly repetitive motions
- frequent or continuous excessive deviation from neutral postures
- frequent or sustained and excessive exertion of pressure

Manual input devices such as keyboards, mice, pucks, trackballs, and lightpens typically result in arm pronation and wrist extension. Solutions involve redesign of input devices, operator training, and the design of appropriate workstations. New ergonomic standards have attempted to address this problem by

- requiring input devices to be designed so that they can be continuously used in neutral postures and without excessive biomechanical load
- requiring workstation designs that allow neutral postures
- recommending appropriate user training

However, since most input devices on the current market do not currently meet these requirements, the future should bring many novel designs.

There are now many products on the market that claim to be "ergonomically designed" and solve a range of VDT problems from carpal tunnel syndrome to radiation exposure. Unfortunately, many of these products not only do not solve these problems, but they may create others. For example, some of the new split keyboard designs are designed to reduce ulnar deviation, extension, and arm pronation by raising the center of the two halves of the keyboard. However, the angle of tilt of some of these keyboards is often greater than 15°, which eliminates wrist ulnar deviation but causes radial deviation. Also, many mice are shaped to conform to the hand but the button size and location result in excessive finger separation and their height causes excessive hand extension.

Designing for the Disabled

Products should be designed for easy use by all anticipated users including the disabled. At a minimum, products should be easily used by at least 95 percent of anticipated users—including persons who are disabled.

In the U.S., the passage of the Americans with Disabilities Act (see Chapter 6) has significantly influenced special designs for the disabled. Until recently, most high technology manufacturers and distributors had typically not made efforts to design products easily usable by individuals with disabilities. Only two of the major U.S. computer manufacturers (IBM and Apple) have development divisions dedicated to designing and modifying their products to accommodate the disabled.

Office work is easier for disabled users than work requiring heavy labor. Companies producing products for office work should thus attempt to design their products for the disabled as well as users without disabilities. Future markets will require product features that enhance use by all users including the disabled. These user features include improved visibility of labeling (see section on Control Panels) for users with low vision and improved access and use for people with muscular disorders. These improvements will result in benefiting all users. For example, poor visibility of labels on control panels is a problem for most users.

Some of the problems of product accessibility for the disabled can be addressed by hardware design, others by software design. For example, 8–10 percent of males and 0.5 percent females have color vision deficiencies—the majority of whom are unable to discriminate red from green. Software designers can create default color palettes using only colors that can be discriminated by all users. For example, cyan can be used instead of green and magenta instead of red.

Displays

Flat panel displays should be produced that are at least the same size as, offer as good or better resolution than, and are cost competitive with cathode ray tubes (CRTs).

A challenge for product developers is to create products that can accommodate the limitations imposed by modern offices including the limited space in modular/cluster workstations. There is an increasing demand for flat panel displays because they occupy less work surface space and they can be easily moved. A major problem with flat panel displays is that their cost continues to prohibit mass purchase of large (that is, 12–17 inch) displays. In order to display the same quantity of text on a cost comparative flat panel display as is displayed on a standard 12-inch CRT, the text on the flat panel display will have to be smaller than is ergonomically acceptable because the display is smaller. Viewing text smaller than ergonomic standards require can result in visual and musculoskeletal problems as users squint and bend forward to focus on small characters. In addition, many of the flat panel technologies do not yet allow visibility at as wide a viewing angle as CRTs.

An additional challenge to display designers is the inclusion of automatic brightness and color contrast features. This feature would greatly enhance visibility of all displays particularly for lap top systems, which are used in a variety of light conditions.

As described in previous chapters, the increased interest in ergonomic standards is a result of concerns regarding the effect new technologies have on worker health and safety. The concerns center around vision, musculoskeletal functions, physiological processes, and reproduction. Although most research has not shown a causal relationship between display use and visual and reproductive disorders, studies on these issues continue, and results often show high correlation between these disorders and display use. For manufacturers to be competitive in the future, they should offer flat panel displays with large screens. In addition, all displays should have low radiation emissions.

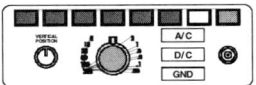 Control Panels

Labels on control panels should be larger than traditionally recommended and have good contrast.

The labeling on most product control panels typically does not meet ergonomic standards for contrast and size. The problem becomes particularly apparent in offices with inconsistent, or low ambient lighting conditions and with users who are older—an increasingly common situation. Older individuals require large contrasts to clearly see edges and other image details. Low contrast labels such as light gray characters on medium gray product covers, which are common on panels on many computer products, are not sufficient for good visibility.

Studies demonstrate that high resolution and contrast result in images that are easy to see and in characters that are easy to read. Most standards for the design of control panel labels were based on studies of relatively young to early middle-aged adults with good vision. In addition, standards and recommendations for label size and contrast are often not followed in order to conserve space or because of aesthetic reasons. Characters are thus often too small and of insufficient contrast to be seen at typical viewing distances and viewing angles. To address these problems and accommodate visual abilities of older users, characters need to be at least 20 percent larger than those traditionally recommended. Large characters on labels increase visibility at a variety of viewing distances and for a wider viewing angle. This improves the usability of equipment for a variety of users in different work conditions and environments.

In addition to the problem of character size, many labels on office product control panels are in an inappropriate polarity, that is dark images on a lighter background. Research has shown that light images on dark backgrounds are easier to see than dark characters on light backgrounds, particularly for illuminated labels (see Chapter 2). In addition, many labels on control panels have insufficient contrast (that is, less than 10:1). This low contrast is common for white labels on a light gray background and black labels on a medium gray background. Designers of control panel labels should use bold text labels that are lighter than their background with sufficient contrast to ensure good legibility at anticipated viewing distances and angles.

Icons and Fonts

Ergonomic principles derived from studies of visual perception should be used in the design and selection of icon and font designs. Likewise, the design of icons and fonts should be validated by legibility data and testing. Ergonomically validated icons and fonts should be used as default fonts.

Font and icon designs are typically based on aesthetics and graphic design preferences. The result is that many icons in tool bars and fonts are difficult to identify and interpret (see Figure 10–1).

Ergonomic standards for icons have recently been initiated (see Chapters 1 and 5). However, there are no activities to create ergonomic standards specifically for legibility and readability of font designs.

Some attempts have been made to create metrics to test the usability of icons. These tests include evaluations of visual and cognitive aspects of icons, such as their detection, identification, differentiation, and interpretation. More research is needed on the ergonomic design and testing of icons and fonts.

Easy to read   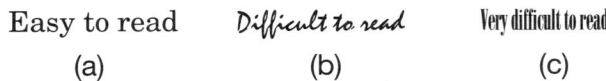   Very difficult to read

(a)   (b)   (c)

Figure 10–1 Legibility of software bundled fonts. All three fonts are the same pitch size and the default light face. The legibility of font (c) is most difficult because of the inappropriate height-to-width ratio and close spacing of the letters.

Product Weight and Size

Products should be small and light weight without increasing complexity.

The increasing mobility of users requires products that are easy to transport and install. Computer workstations, printers, and plotters were not originally built to be frequently moved but are often transported between offices and home. Dissatisfaction with heavy, large products has created a marketing opportunity for small, light weight computer products. The Japanese are well aware of this opportunity (see Chapter 7).

These new light, high technology products allow users more flexibility in arranging components in their workstations. Computers that 15 years ago occupied an entire desk top can now be held in a user's hand; printers have shrunk from six cubic feet to less than one cubic foot. Unfortunately, miniaturization has often resulted in condensing functions to a few controls that have layers of complicated menus. Reducing the size of a product should not reduce required functionality nor increase user interface complexity.

Cable Management

Products need to have a better method of connecting computer components so that the number of cables is reduced, or at least combined in a more efficient manner.

The typical computer workstation has seven cable connections:

1. the keyboard to the processor
2. the mouse to the processor
3. the display to the processor
4. the display to the power supply
5. the printer to the processor
6. the printer to the power supply
7. the processor to the power supply

The addition of peripheral products (like scanners, plotters, tape drives, speakers, and CD ROMs) add more and more cables that exacerbate that problem. Cables become tangled, interfere with portability, limit location flexibility, and produce safety hazards. The need for compact and clever cable management is becoming increasingly more important.

Heat and Air Circulation

Future computer products should generate significantly less heat than current systems or generate no heat at all. Offices and modular workstations need to incorporate effective air ventilation and air purification systems.

Many offices in older buildings are not air conditioned, especially in countries where older buildings are the norm. In addition, most offices and homes are not designed to accommodate the amount of heat generated by computer equipment. The result is that these environments can quickly, but subtly, become excessively warm and dry. In addition, the fans from computer terminal products often produce air currents that are easily perceptible and annoying to users and adjacent workers. These conditions can result in visual and general discomfort, skin problems, and a reduction in cognitive functioning and user performance. For example, low humidity and excessive air circulation increase eye dryness especially for users who wear contact lenses and for users who blink less often as they intently concentrate on computer screens. As workplaces become substantially warmer than 75°F, concentration becomes more difficult, and cognitive performance becomes easily degraded. The high incidence of skin problems that often occurs when using VDTs in dry offices has been the subject of numerous papers published and presented at international conferences for the last decade. However, when cluster workstations are used, employee health and comfort improve, and worker absenteeism can be reduced when air purifiers are incorporated into the workstations.

Workstations

Evaluate the appropriateness and selectively use modular workstations.

The trend in American office designs for the last twenty years has been an increase and often inappropriate use of modular, cluster workstations. Research has shown that open-office, cluster workstations are not appropriate for all types of work and environments. They are best and most effective for individuals that work in teams. Private offices are best for workers that require privacy and a quiet work environment whether for security or concentration.

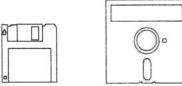 Software User Interface Design

Companies need to change and improve substantially their software user interfaces. They need to create interfaces that are more intuitive to learn and use, to improve the explanation of errors, and provide more understandable error recovery procedures. On-line Help could be more effective, particularly for novice users, if it was an interactive tutorial which included animated and multimedia demonstrations.

Millions of dollars are spent monthly on telephone and E-mail technical support for software problems. This is evidence that the design of the user interface and on-line Help is seriously deficient. Most on-line Help is more like an encyclopedia than a tutorial or error-recovery assistant. The majority of on-line Help programs are monologues rather than helpful dialogues with the user. Few on-line Helps use graphics or animation; almost all are text dominated in a world where few users have the time or patience to scroll through pages of Help text. Since ergonomic research has demonstrated faster and more thorough learning when images are added to text, the addition of graphics to an on-line Help would also improve learning.

Because the cost and function of computer software products from different manufactures are now so similar, the few ways a company can differentiate its products are by the appearance and interactions ("look" and "feel") of their user interfaces. To achieve this goal, the majority of software corporations need to improve their products substantially. For example, very few mass users of Windows operating systems understand and can correctly modify basic system files such as *autoexec.bat, config.sys, system.ini* and *win.ini*. These commands appear so complicated and their modifications so serious that most users do not even bother to try to learn these procedures. UNIX is another operating system that is well known for being difficult to learn and use and is thus primarily used by highly trained people. However, studies have shown that, in spite of their advanced knowledge, UNIX users generally employ 10 percent of UNIX commands 90 percent of the time. These examples demonstrate the need for easier-to-use systems.

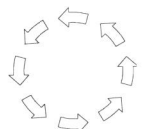 **Process Strategies**

Development and Marketing

Products should be evaluated throughout the design process to determine if they meet minimum ergonomic design principles and standards. Human factors engineering/ergonomics needs to be integrated into the entire product development and marketing process. Companies need to have sufficient flexibility to adapt their product development processes to accommodate new ergonomic findings, standards, and regulations.

An ergonomic design process should include developing and using the following:

- a description of user characteristics (including disabled users)
- a description of user tasks
- a description of environments in which the products will be used
- a set of minimum design requirements based on ergonomic principles and standards (see Chapter 8)
- usability goal specifications and checkpoints
- usability tests appropriate for application validation (see Chapter 9)

Some companies are integrating human factors engineering/ergonomics into their development process and product design. These companies have become internationally recognized for the ease of use/ergonomic design of their products. They realize that ergonomic design is an essential component of their product engineering. Moreover, they have developed a clear understanding of their product users, user needs and user product expectations. These companies successfully translate this information into innovative products that meet ergonomic principles and standards.

Corporations need to integrate ergonomics into product development and marketing. In order for this integration to be taken seriously, it needs to be supported by executive management, given high priority and become an integrated part of the development process. Only a few corporations have top executives who are familiar with the ergonomic activities in their company and place ergonomics/user interface among their companies top ten priorities.

Some companies (such as Hewlett-Packard and many Japanese companies, see Chapter 7) have made product usability one of their primary corporate objectives. Their ability to achieve this goal is directly dependent on the quality and value of their products as perceived by their users. Good interface design should be a primary goal for all product development organizations. Reaching this objective can enable companies to achieve standardization while retaining product differentiation, high value, and uniqueness in the marketplace. It is through this process that ergonomic specifications can be configured into economically viable products, services, and work environments. Good ergonomics, from a commercial standpoint, will satisfy corporate objectives of effective, safe, easy and comfortable-to-use products while building profits.

Design and Testing

Product designs based on a program of partial ergonomic analysis (see Chapter 9), cannot truly address or solve ergonomic problems. While such products may result in partial or short-term solutions that gain short-term profits, they often loose their competitive edge in the long run. Design without appropriate and sufficient ergonomic testing often occurs because of the following reasons:

- Companies do not want to invest in the cost of testing.
- Many companies believe that focus groups are sufficient to determine acceptance, appropriateness, and thus success of product designs.
- Designers or do not believe their designs need to be validated.
- Designers are afraid testing will demonstrate a weakness in their designs.
- Developers do not believe there is sufficient time for testing.
- Developers are afraid testing will show that design changes are necessary and will lengthen development time and thus time-to-market.
- Manufacturers are afraid testing will demonstrate a serious problem that might prohibit marketing of the product or result in recall of products already on the market.
- Manufacturers do not have personnel appropriately trained in usability testing and data analysis techniques or have appropriate testing facilities.

ISO 9000

The standard dedicated to determining if a company has the capabilities and processes to provide quality products and services is ISO 9000. Although ISO 9000 is not an ergonomic standard, ergonomics can assist in obtaining compliance with it. In addition, meeting ISO 9241 requirements can help meet the process requirements of ISO 9000.

ISO 9000 Requirements

The ISO 9000 standard was initially developed as a specification that would eventually be mandated for suppliers wishing to do business in the European Union (EU). Although ISO 9000 is currently a standard, it is expected to become a legal requirement in the EU. By 1992, 51 countries had adopted ISO 9000 as a national standard. However, no other national or multinational organization except the EU appears to be converting it into a legal requirement.

There are five parts to ISO 9000 (see Table 10–1). Parts 1–3 describe quality system models and are of varying stringency for use in different applications; Parts 0 and 4 are guidelines for the use of the standard.

ISO 9000 Certification

To obtain ISO 9000 certification, a company must keep detailed accounts of its procedures in engineering, manufacturing and services and demonstrate that it has a recordable method to track the activities that occur in these functions. Independent auditors are required to conduct an evaluation of a company's site quality management system. The auditor uses the ISO 9000 standard as a benchmark. Audits must occur at least once every three years.

All the activities that occur in the design cycle of a product are required to be thoroughly described in a company's documentation. The process description should include the following details:

Table 10–1 ISO 9000 Parts

Part No.	Title	Contents
9000	*Quality Management and Quality Assurance Standards—Guidelines for Selection and Use*	Explanation of fundamental quality concepts Definition of key terms Guidance in selecting, using and tailoring ISO 9001, ISO 9002, and ISO 9003
9001	*Quality Systems—Model for Quality Assurance in Design/Development, Production, Installation and Servicing*	Design capabilities Development capabilities Servicing capabilities
9002	*Quality Systems—Model for Quality Assurance in Production and Installation*	Prevention of problems Detection of problems Correction of problems
9003	*Quality Systems—Model for Quality Assurance in Final Inspection and Test*	Detection of problems Control of problems
9004	*Quality Management and Quality System Elements—Guidelines.* Used as a tool for internal and external auditing	Supplier guidance for: ⇒ developing and implementing a quality system ⇒ determining the extent to which each quality system element is applicable

- specific product-design and manufacturing activities
- specified points in the process when each activity occurs
- methods of assessing the quality of product design and production
- qualifications of the product designers, assemblers, and evaluators
- verification that pre-determined specifications and standards are incorporated in designs

ISO 9000 Compliance

In order to comply with Parts 1–3, a company must demonstrate the following:

- Engineering, manufacturing, and service quality assurance protocols exist and are routinely used.

- Quality measurement protocols are valid.
- Measuring and testing equipment is regularly calibrated.
- Method for identifying and tracing products is in place.
- Process for handling, storing, packaging, and delivering of products is adequate.
- Process for inspecting, testing, and dealing with products that do not conform to inspection standards is adequate.
- Internal audit system exists.
- Compliance to quality standards can be correctly demonstrated.
- Quality training program is appropriate and is given to appropriate staff.
- Employees are appropriately trained for their jobs.

ISO 9000 Audits

ISO 9000 audits are conducted by individuals or agencies that typically have different interpretations of the ISO 9000 standard requirements. Agencies often use their own checklists, auditing questions, and interpretations of the resulting data. This is because there is no standard, or required, compliance metrics for ISO 9000 and because many requirements of the 9000 standard use vague compliance descriptions such as "where appropriate" or "consideration should be given to." Although the requirements are clearly defined, the way in which the requirements are to be met is vague and thus left open to interpretation.

Implications for Design

ISO 9000 has the following implications for user interface designers:

- A way of assessing the quality of the design and ergonomics needs to be established.
- Quality assessment protocol(s) (including usability testing) should be established and validated.
- Design methods should be periodically calibrated.
- A way to identify and trace the progression of designs should exist.

- An internal audit of a design with appropriate documentation is needed.
- Compliance with appropriate standards (including ergonomic standards) must be demonstrated in designs.
- Design iterations and evaluations must be documented.
- A program to train designers in ISO protocol(s) must be available, and designers must participate in the training programs.
- Designers must have appropriate
  - training for the type of design for which they are responsible
  - knowledge of materials and fabrication processes
  - appropriate knowledge of relevant fields like biomechanics, perception, and cognition

The Role of ISO 9241 in ISO 9000

ISO 9241 is classified as a quality standard. Complying with ISO 9241 can assist in complying with ISO 9000. For example, Part 11 of ISO 9241 requires usability specifications that include a description of user characteristics, use of applications, and environments in which the product will be used and how its usability is to be measured. The information provided by these descriptions can demonstrate to an ISO 9000 auditor that the design and its process has appropriately addressed the user and market requirements of ISO 9001.

The Future and Ergonomic Standards

A number of factors will influence future national and international ergonomic standards. They include:

- influence and expansion of European standards and legislation
- user and labor demands
- research of health issues related to the use of high technology products and environments

- growing interest in and funding of ergonomic research
- interest in standardizing new technologies, design features, applications, and usability metrics
- increase of ergonomics as a major purchase factor and product discriminator
- increase in consideration and use of ergonomics in legislation, labor negotiations, and litigation
- interest in ergonomic certification and the need for certification methods and laboratories
- demand for ergonomic/user quantitative and qualitative metrics
- use of data to validate and substantiate designs
- demand for user interface features by users with special requirements
- use of artificial intelligence/expert systems as a user interface design and evaluator aid
- use of systemic, rather than a component, approach to ergonomic design and testing
- creation of consumer product ergonomic standards
- creation of medical product standards
- interest in, and development of, on-line trainers and performance evaluators using multi-media
- ergonomic research and standards on virtual reality, voice interfaces, and multi-media

Astute product designers and manufacturers will proactively respond to these trends and support and participate in ergonomic activities.

Summary

The increase in ergonomic standards has significantly impacted product design and competition. Development organizations are beginning to incorporate ergonomics into their product life cycles; ergonomic design features are being incorporated into products and advertising media. Manufacturers are realizing that these efforts can reduce product development time and cost and increase profits. They are also becoming increasingly aware that without these efforts, they loose market share.

The concerns of health and safety associated with use of computer and other high technology products are expected to continue and be evidenced in increasing ergonomic research, standards, legislation, product development, measurement metrics, and industrial competition. The impact of ergonomics on standards, regulations, and high technology products will become increasingly more salient. In the future, manufacturers' ability to produce, market, and sell products will to a large extent depend on their ability to

- understand ergonomic issues of new technologies
- produce technologies that meet ergonomic requirements
- provide data that substantiates product designs

Only the future will decide the ultimate impact of ergonomics on product standards, design, and use. The integration of ergonomics into product design and development as well as marketing processes may someday be the deciding factor in the success or failure of products as well as their manufacturers.

Glossaries

Standards Agency Glossary

ANSI	American National Standards Institute
ASA	Acoustical Society of America
ASTM	American Society for Testing and Materials
BSI	British Standards Insitution
BSR	Board of Standards Review
CBEMA	Computer Business Equipment Manufacturers Association (see ITI)
CCITT	International Telegraph and Telephone Consultative Committee
CEC	Commission of the European Communities
CEN	European Committee for Standardization
CENELEC	European Committee for Electrotechnical Standardization
CHI	Computer Human Interaction
CIE	Commission Internationale de l'Eclairage (International Commission of Lighting)
COMECON	[European] Council for Mutual Economic Assistance

CSA	Canadian Standards Association
DIN	Deutsches Institut fur Normung (German Institute for Standardization)
DOD	[U.S.] Department of Defense
DOT	[U.S.] Department of Transportation
ECMA	European Computer Manufacturers Association
EEA	Agreement on the European Economic Area
EFTA	European Free Trade Association
EOTC	European Organization for Testing and Certification
ESPRIT	European Strategic Program for Research Development in Information Technology
ETSI	European Telecommunications Standards Institute
EU	European Union
FIET	International Federation of Commercial Clerical and Technical Employees
GSA	General Services Administration
HFS	Human Factors Society (former name of current Human Factors and Ergonomic Society)
HFES	Human Factors and Ergonomic Society
HSC	[United Kingdom] Health and Safety Commission
HSE	[United Kingdom] Health and Safety Executive
ICSID	International Council of Societies of Industrial Design
IEA	International Ergonomics Association
IEC	International Electrical Commission
IEEE	Institute of Electrical and Electronics Engineers
ILO	International Labor Organization
ITI	Information Technology Industry Council (formerly CBEMA)
ISO	International Organization of Standardization
JIS	Japan Insitute of Standardization
JTC	Joint Technical Committee

MITI	Japanese Ministry of International Trade and Industry
MPR	[Swedish] Board for Measurement and Testing
NIST	National Institute of Standards and Technology
OSHA	Occupational Safety and Health Association
SIGCHI	Special Interest Group on Computer and Human Interaction
TCA	[German] Trade Cooperative Association
TCO	[Swedish] Central Organization of Salaried Employees
TUC	Trades Union Congress
WHO	World Health Organization

Standards Glossary

ADA	American with Disabilities Act
ANSI HFS 100/1988	American National Standard for Human Factors, Engineering of Visual Display Terminal Workstations
ANSI Z-365	American National Standards Institute, Control of Work-Related Cumulative Trauma Disorders
ANOHSC	Australian National Health and Safety Commission, Prevention and Management of Occupational Overuse Syndrome
ANOHSC	Australian National Health and Safety Commission, Screen Based Workstations
BS 7179	British Standards Institution, Ergonomic of design and use of visual display terminals (VDTs) in offices
CAN/CSA-Z412-M89	Canadian Standards Association, Office Ergonomics
CEN PIEN TC 122	Preliminary Release of European Norm, Basic List of Definitions of Human Body Dimensions for Technical Design

CIE 15.2	Commission Internationale de l'Eclairage, Colorimetry, Publication No. CIE 15.2, Central Bureau of the CIE, Vienna, 1986
CIE 17.4/IEC 845	Commission Internationale de l'Eclairage, International Lighting Vocabulary, Central Bureau of the CIE, Vienna, 1989
CIE 63	Commission Internationale de l'Eclairage, The spectroradiometric measurement of light sources
CIE/ISO 10526	Commission Internationale de l'Eclairage/ International Organization of Standardization, Colorimetric Illuminants, Central Bureau of the CIE, Vienna, Austria, 1991
DIN 66234	German VDT standard, Visual Display Unit Work Stations

        Part 1 Geometric design of characters
        Part 2 Perceptibility of characters on screens
        Part 3 Grouping and formatting of data
        Part 4 [not assigned]
        Part 5 Coding of infomation
        Part 6 Workstation design
        Part 7 Ergonomic design of environmental lighting and envrionment
        Part 8 Principles of man-machine dialogue design
        Part 9 Measuring methods
        Part 10 Minimum information to be specified

EC [EU] Directive 89/391	European Union Directive, Health and Safety Directive
EC [EU] Directive 90/270	European Union Directive, Minimum Safety and Health Requirements for Work with Display Screen Equipment

GSA 506	United States General Services Administration, Access to Information Technology by Users with Disabilities.
HSE L21	United Kingdom Health and Safety Commission regulations, Management of Health and Safety at Work
HSE L26	United Kingdom Health and Safety Executive regulations, Display Screen Equipment Work
ISO 7000	Insternational Organization for Standardization, Graphical Symbols for Use on Equipment—Index and Symbols
ISO 9000	International Organization of Standardization, Quality Assurance Standard.
IEC 73	International Electrotechnical Commission, Colors of Indicator Lights and Pushbuttons, 1965
IEC 950	Safety of Information Technology Equipment, including Electrical Business Equipment
ISO 9241	International Organization of Standardization, Ergonomic Requirements for Office Work with Visual Display Terminals

    Part 1   Introduction
    Part 2   Task Requirements
    Part 3   Display Requirements
    Part 4   Keyboard Requirements
    Part 5   Workstation Layout and Postural Requirements
    Part 6   Environmental Requirements
    Part 7   Display Requirements with Reflections
    Part 8   Requirements for Displayed Color
    Part 9   Requirements for Non-keyboard Input Devices

	Part 10 Dialogue Principles
	Part 11 Guidance on Usability
	Part 12 Presentation of Information
	Part 13 User Guidance
	Part 14 Menu Dialogues
	Part 15 Command Dialogues
	Part 16 Direct Manipulation Dialogues
	Part 17 Form Filling Dialogues
JIS 6041	Japanese Standards Institute, CRT Display and Keyboard Units for Business Use
JIS Z 8500	Japanese Standards Institute, Ergonomics —Anthropometric and biomechanic measurement
JIS Z 8513	Japanese Standards Institute, Visual display terminals
MPR 2	Swedish Board for Measurement and Testing, Test Method for Visual Display Units
ZH 1/635	German regulation, Equipment Safety Law
ZH 1/618	German regulation, Display Workplaces in the Office Sector

Standards Terms

DAD	Draft Addendum
DAM	Draft Amendment
DIS	Draft International Standard
DP	Draft Proposal
DTR	Draft Technical Report
EN	European Norm
IS	International Standard
PREN	Preliminary Release of European Norm
SC	Subcommittee
TAG	Technical Advisory Group
TC	Technical Committee

TR   Technical Report
WG   Working Group
WI   Work Item

Hardware User Interface Glossary

bar code	A series of variable-width, parallel lines representing information about an item
bar code reader	A hand-held optical device that transmits spatial forms on coded labels, bar codes and characters to a computer; also known as a hand-held scanner
brightness	Attribute of a visual sensation in which an area appears to emit more or less light
button	A mechanical object integrated into an input device that when pressed provides input to the computer
cathode ray tube	An electron tube which creates lights by focusing electron beams on the back of a phosphor coated computer screen
click	Audible or tactile indication of the actuation point of a key or button
color detection	The perception of a color sensation
color difference calculation	The calculated difference between two color stimuli, defined as the Euclidean distance between the points representing them in a color space
color discrimination	Detection of color difference between visual stimuli
color identification	The ability to name a color
color interpretation	Association of a particular color to a meaning or function.
contrast ratio	The ratio between the higher (L_H) and lower (L_L) luminances that define a feature to be detected and measured and calculated by:

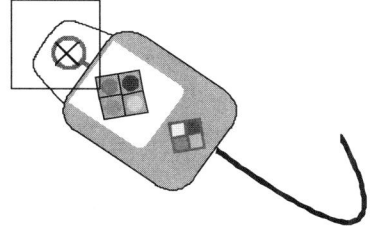

Figure G-1 Puck with reticule (in square) showing cross aligners

$$CR = \frac{L_H}{L_L}$$

convergence	The exact intersection of electron beams on a color CRT at a specific point on the plane of its phosphor screen
cross-aligners	Orthogonal lines (see Figure G-1) in a reticule used to align an input device visually with an image; also known as cross-hairs
cursor	The symbol on the computer screen that indicates a pointer or location of the next input
default color set	A predetermined group of colors assigned by the software application or operating system
defective color vision	Anomaly of vision in which there is a reduced ability to discriminate some or all colors
depth-of-field	The distance within which all images are in focus
design viewing distance	The distance or range of distances between the screen and the operator's eyes for which the display is designed to be viewed
design working position	The position and posture for which a design is intended to be used
drag	Controlling the movement of a selected image across a display screen with an input device

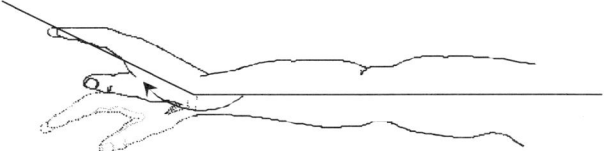

Figure G–2 Hand extension

extension (hand) — An increase in the angle between two adjacent bones; hand extension is the movement of the hand in the dorsal (as opposed to the palmar) direction (see Figure G–2)

feedback — An indication of the results of a user action. Input device feedback refers to a tactile, auditory, or visual indication that an activation has occurred. Display feedback refers to a change on the display resulting from an input device movement or activation.

finger abduction — Separation of the fingers (see Figure G–3)

flexion (hand) — A decrease in the angle between two adjacent bones; hand flexion is the movement of the hand in the palmar direction (see Figure G–4)

freehand input — Manual input without the use of guiding aids

gain — The ratio of input device movement to cursor movement

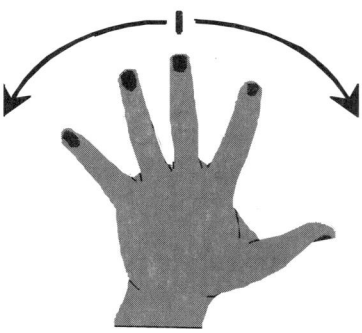

Figure G–3 Finger abduction

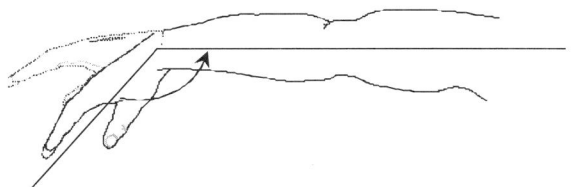

Figure G–4 Hand flexion

goniometer	Special protractor for measuring joint angle
graphic digitizer	An input device that converts graphic data into binary inputs
graphic tablet	A tablet which senses the absolute position of a pointing device on its surface
hue	A visual sensation of a specific color
input device	A user-controlled device that transmits information to a computer
joystick	A lever mounted in a fixed base used to control the movement of objects displayed on a computer screen. An *isometric joystick* is a pressure-sensitive joystick whose handle does not move when pushed. A *displacement joystick* moves in the direction of force and the cursor moves in proportion to the distance the lever is displaced
lateral hand deviation	Position of the hand either to the left or to the right of neutral (see Figure G–5)
light-pen	A pen-like input device (see Figure G–6) that contains a light detector or photocell

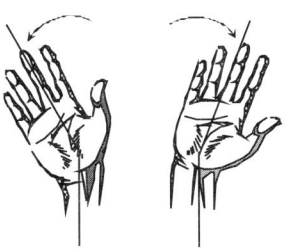

Figure G–5 Lateral hand deviation

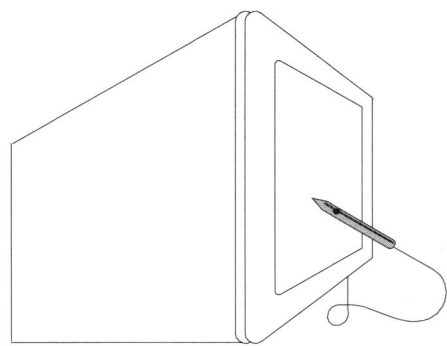

Figure G–6 Light pen and screen

	and generates position selection and information when pointed at the display
mouse	A hand-controlled device that typically fits the palm (or fingers) and is moved on a flat surface in an x and y direction; buttons are typically located on the top or front surface of the mouse (see Figure G–7) and are used for input or image manipulation
neutral hand/arm position	Direct alignment of the hand with the lower arm
palm/wrist rest	A surface on which a user may place the palm of the hand/wrist when using an input device
parallax	The perceived displacement between the registration and the target due to the dis-

Figure G–7 Mice

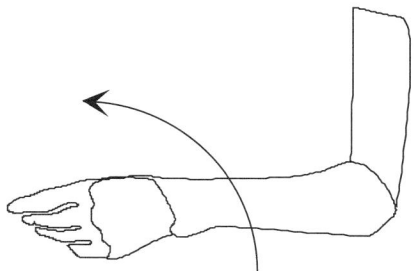

Figure G–8 Arm pronation

	tance between the touch sensor and the target
point and click	To position a cursor over an object displayed on a screen (point) and press a button (click) or pointing device to select it
pointer	A symbol indicating input or selection position on a computer screen whose movement is controlled by the user (see also *cursor*)
pointing	The process of using an input device to indicate locations on a display
pronation (arm)	Inward (medial) rotation of the forearm (see Figure G–8)
puck	A hand-held device similar to a mouse, with a reticule (e.g., cross-hair) view finder (see Figure G–9) typically used with a tablet (see Figure G–13)

Figure G–9 Puck

radial hand-deviation	Bending the hand at the wrist in the direction of the thumb (see Figure G–10)
resolution	The smallest detectable movement or activation force of an *input* device; the smallest cursor displacement of an *output* device
reticle	A system of lines, dot, cross-hairs or wires in the focus of a lens, such as in a puck (see puck).
RSI	repetitive strain injury
saturated color	A color with a chromatic purity of one (1)
saturation	Visual sensation of colorfulness of an area in proportion to its brightness
scanner	An input device that encodes images or text from paper and transmits them to a computer for viewing or manipulation on a screen
selecting	The process of choosing one or more items on a display
simple graphics	Computer-generated graphs, charts, icons, and pictures composed of lines or area-fill that are not continuous shades or have the appearance of a photograph
stereopsis	Binocular visual perception of depth or three dimensional space

Figure G–10   Radial hand-deviation

Figure G–11 Stylus and tablet

stylus
: A pen-shaped pointing device that when placed on a display or graphics tablet (see Figure G–11) can be used to draw images displayed on a computer screen; select objects typically via the depression of the stylus tip; one or more small buttons located along the side of the stylus

supination
: Lateral (outward) rotation of the forearm that brings the palm of the hand upward (see Figure G–12)

tablet
: A flat surface on which movement of a stylus (see Figure G–13), puck, or mouse generates position and selection information of images on a display screen

tablet overlay
: A thin template on the surface of a tablet (see Figure G–14) used to indicate the configuration of functions available to the user

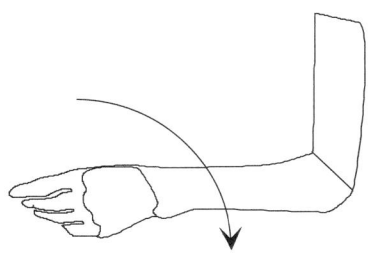

Figure G–12 Arm suppination

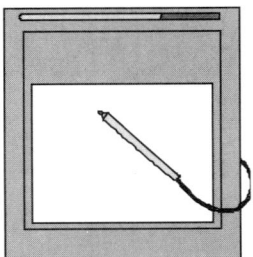

Figure G–13 Tablet with stylus

tactile feedback	An indication of the results of a user action transmitted through the sense of touch
thumbwheel	A rotating wheel, with a portion of its edge exposed for manipulation by the fingers, used to control a single variable
touch-sensitive screen	A device that produces an input signal containing position and/or selection information in response to touching or moving a finger on a display
tracing	Copying an image by moving an input device over the image lines and/or shape
tracking	Moving a cursor or predefined symbol across the surface of a display screen in order to follow a target
trackball	A ball in a fixed housing (see Figure G–15) that can be rolled in any direction by the fingers to control cursor movement

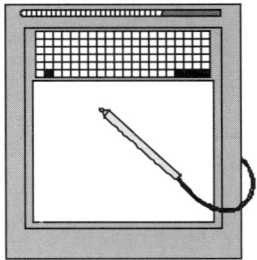

Figure G–14 Tablet with overlay

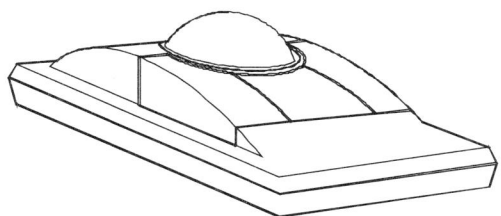

Figure G–15 Trackball input unit

translation
: The movement of a cursor or input device from point to point along a path or straight line

ulnar hand deviation
: Moving of the hand at the wrist in the direction of the little finger (Figure G–16)

visual adaptation
: The process by which the state of the visual system is modified by exposure to stimuli that may have various luminances or spectral distributions

workplace
: The area (including room lighting, noise, temperature, etc.) immediately surrounding the worksation

workstation
: A single location within a workplace at which instrumentation or equipment is located and at which a worker remains for extended periods of time to perform functions such as information control, monitoring, and processing

wrist rest
: See *palm rest*

Figure G–16 Ulnar hand deviation

Software-User Interface Glossary

active window	A currently selected window
check box	Control that allows user to switch a single attribute to "on" or "off"
choice list	Temporarily displayed list containing a variable number of items from which a user can select
command button	Rectangular button on a keyboard that when pressed carries out the specific action that is indicated by the label on the button
control	Allows user to interact with data and objects on the user interface
control menu	Menu that allows user manipulation of a window
control menu box	Box located in upper corner of each window or dialogue box
default button	Command button in a dialogue box with a thicker border that is selected by default and can be activated by depression of enter key
dialog box	Special type of window that requests or provides information
drop down list	Variation of a choice list with single selection. The current selection is displayed in a one item list and other choices are displayed on user requests
HCI	Human Computer Interaction
icon	A graphic representation of an application item or action
keyword	Any word used as a command
list box	Dialog box component that contains a list of available choices
manipulation	Any user interaction with a graphical component of the user interface that includes selecting, positioning, orienting, qualifying, pathing, or sizing in order to achieve the goal of a task

Software-User Interface Glossary

menu bar	A horizontal list of the major menu options displayed at the top of a window
metaphor	Analogy to concepts that are already familiar to the user and from which the user can derive the systems use and behavior
object	An entity with can be manipulated by the user during the dialogue
pointer	A graphical symbol moved on a screen according to manipulations of a pointing device (mouse, puck, stylus)
pop-up menu	Menu displayed next to an object or set of objects after object selection
pull-down menu	A vertical list of related menu options that extends down from an activated menu bar option
push button	A rectangular input-sensitive area on the screen that a user can select in order to start a dialogue function
positioning	Moving or dragging an object or its graphical representation to another area on the screen
radio button	A control that allows the user to change related attributes or parameters of objects
scroll bar	An on-screen control that allows the viewing of material in a file that extends beyond the display area of a window; permits both vertical and horizontal movement of file material
scroll box (slider)	A small box that slides either vertically or horizontally in the scroll bar indicating the relative position of the file image being viewed
title bar	Part of a window or dialog box that shows either the name of the application that is running or the name of the dialog box
UI	User interface
window	An independently controllable area of a screen, usually rectangular and usually delimited by a border

Appendix

National Standard Agencies & ISO/TC 159 Participants

Country	Standards Agency Abbreviation	Participant	Observer
Albania	KCSA		
Algeria	INAPI		✓
Argentina	IRAM		✓
Australia	SAA	✓	
Austria	ON	✓	
Bangladesh	BST		
Belgium	IBN	✓	
Brazil	ABNT		✓
Bulgaria	BDS		
Canada	SC	✓	
Chile	CSBS		✓
China	CSBS	✓	
Colombia	ICONTEC		✓
Cuba	NC		✓
Cyprus	CYS		

Appendix

Country	Standards Agency Abbreviation	Participant	Observer
Czechoslovakia	CSN	✓	
Denmark	DS		✓
Eqypt	EOS		✓
Ethiopia	ESI		
Finland	SFS	✓	
France	AFNOR	✓	
Germany	DIN	✓	
Ghana	GSB	✓	
Greece	ELOT		✓
Hungary	MSZH	✓	
India	BIS		✓
Indonesia	DSN		✓
Iran	ISIRI		
Iraq	COSQC	✓	
Ireland	NSAI		✓
Israel	SII		✓
Italy	UNI	✓	✓
Ivory Coast	DINT		✓
Jamaica	JBS		
Japan	JISC	✓	
Kenya	KEBS		
Korea, North	CSK		
Korea, South	KBS		✓
Libya	LYSSO		
Malaysia	SIRIM		
Mexico	DGN		✓
Mongolia	MSC		
Netherlands	NNI	✓	
New Zealand	SANZ		
Nigeria	SON		
Norway	NSF	✓	
Pakistan	PSI		✓
Papua New Guinea	PNGS		
Peru	ITINTEC		
Philippines	BPS		
Poland	PKNMJ	✓	

Country	Standards Agency Abbreviation	Participant	Observer
Romania	IRS		✓
Saudi Arabia	SASO		
Singapore	SISIR		✓
South Africa	SABS		✓
Spain	AENOR	✓	
Sri Lanka	SLSI		
Sudan	SSD		✓
Sweden	SSD	✓	
Switzerland	SNV		✓
Syria	SASMO		
Tanzania	TBS		
Thailand	TISI		✓
Trinidad & Tobago	TTBS		✓
Tunisia	INNORPI		✓
Turkey	TSE		✓
United Kingdom	BSI	✓	
United States	ANSI	✓	
former USSR	GOST		
Venezuela	COVENIN		✓
Vietnam	TCVN		
Yougoslavia	SZS		✓
Zambia	ZABS		

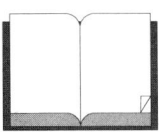# Bibliography

Abernethy, C.N. (1993). Expanding jurisdictions and other facets of human-machine interface IT standards. *Standard View,* 1(1), 9–21.

Abernethy, C.N. (1988). Human-computer interface standards: Origins, organizations, and comment. *International Review of Ergonomics,* 2, 31–54.

Abernethy, C.N., and Akagi, K. (1984). Experimental results do not support some ergonomic standards for computer video terminal design. *Computers and Standards,* 3, 13–141.

Ad Hoc Expert Advisory Committee on Visual Display Terminals. (May 1989). *Report to the department of industrial relations occupations safety and health standards board,* State of California, Department of Industrial Relations, Division of Occupational Safety and Health.

American with Disabilities Act (ADA). (1991). Accessibility guidelines for buildings and facilities, *Federal Register,* 56(144), 35455–35542.

American Psychological Association. (1982). *Ethical principles in the conduct of research with human participants.* Washington, DC.

ANSI. (1986). *American national standard for buildings and facilities—providing accessibility and usability for physically handicapped people, A117.1.* New York.

ANSI. HCI draft standard, *Dialog techniques, Menus* (Chapter 9.2); and *Commands* (Chapter 9.3); *User guidance* (Chapter 10).

ANSI. (1990). *Procedures for synchronization of the national and international standards review and approval process.* American National Standards Institute memorandum ISSB 869.

ANSI. (1993). *Procedures for the development and coordination of american national standards.* New York.

ANSI. (1990). *Procedures for U.S. participation in the international standards activities off the ISO.* New York.

ANSI/HFS. (1988). *American national standard for human factors engineering of visual display terminal workstations,* ANSI/HFES 100. Santa Monica, CA: The Human Factors Society.

APEX. (1979). *Office technology—The trade union response.* London: Association of Professional Executive, Clerical and Computer Staff, Reprocessing Working Party.

Apple Computer. (1992). *Human interface guidelines: The Apple desktop interface.* Menlo Park, CA: Addison-Wesley.

Apple Computer. (1992). *Macintosh human interface guidelines.* Addison-Wesley: Menlo Park, CA.

Arnold, (1984). Ergonomics with bite: New video display standards. *Electronic Business,* 222–223.

Baummer, G. (1988). An international comparison of the prevalence of work-related neck and upper limb disorders among university office workers and its relationship to office work practices. *Ergonomics International 88, Proceedings of the Tenth Congress of the IEA,* Sydney, Australia, 378–380.

Berhqvist, U. (1986). Regulations and recommendations for VDT work—a review. *International Scientific Conference: Work with display units,* 156.

Berns, T. (1985). International standards for VDU design. *Data Processing,* 5, 9.

Berns, T. (1994). The situation in Sweden. *Work with display units,* 2, A3.

Billingsley, P. (1989). The standards factor: A new standards liaison and a new committee. *SIGCHI Bulletin,* 20(4), 14–15.

Billingsley, P. (1989). The standards factor: Are standard user interfaces the solution? *SIGCHI Bulletin,* 21(1), 14–16.

Billingsley, P. (1990). The standards factor: Committee updates. *SIGCHI Bulletin,* 21(4), 16–19.

Billingsley, P. (1990). The standards factor: EC92 and you. *SIGCHI Bulletin,* 21(3), 12–15.

Bondarovskaya, V.M., and Poviakei, N.I. (1994). Complex ecological expertise of VDT as a means for user's health protection in condition of absence of special legislation and standards. *Work with Display Units,* 2, A19.

Branton, T., Leamon, T.B., Metz, B., Ng., T.K., Salzer, H.J., Stewart, T., and, H. (1985). International standardization in the field of ergonomics. *Ergonomics International* 85, G6/1, 655–657.

Brown, C.M. (1988). *Human-computer interface design guidelines.* Norwood, NJ: Ablex Publishing Corporation.

Buhman, K. (1980). Ergonomic and medical requirements in VDU workplaces and corresponding rules within the Federal Republic of Germany. In *Ergonomic aspects of visual display terminals* (pp. 277–281). London: Taylor and Francis.

Bullinger, H-J., Davies, D.G., Faehnrich,K-P., Schackel, B., and Ziegler, J.E. (1986). Research needs and Europe collaboration in human-computer-interaction. *International Scientific Conference, Work with Display Units,* 22–24.

Cakir, A., Hart, D.J., and Stewart, T.F.M. (1980). *Visual display terminals.* London: John Wiley and Sons.

Cakir, A., Reuter, H.J., Scmude, L. von, and Armbruster, A. (1978). *Untersuchungen zur Anpassung von Bildschirmarbeitsplatzen an die physische and psychische Funktionsweise des*

Menschen. Forschungsbericht der Humanisierung des Arberitsleben. Bobb, Der Bundesminister fur Arbeit und Sozialordnung. [*Studies of the accommodation of monitor workplaces on the physical and physiological functions of the human. Study report of humanizing work environment.* Minister of Labor for Work and Social Order.]

CCITT. (1984). *The human interface to visual display terminals. Fascile #2, Recommendations #E-221.* International Telephone and Telegraph Consultative Committee.

CCITT. (1984). *Man-machine language, Fascile V1.13, Recommendations Z.301-Z.341.* International Telephone and Telegraph Consultative Committee.

Central Organization of Salaried Employees. (1986). *VDU work the right way.* Sweden: Bratts Tryckeri AB.

Christie, B. (1985). *Human factors of the user-system interface, A Report on ESPRIT preparatory study.* North Holland.

Chubb, J.R. (1994). Technical and regulatory issues in electromagnetic emission. *SID 94 Digest,* 227–229.

Committee on Education and Labor. (1985). *Oversight of OSHA with respect to VDTs in the workplace,* Serial No. 99-A. Subcommittee on Health and Safety, House of Representatives Congress, First Session, 28 pages.

Conformance Testing and Certification in Information Technology and Telecommunications. (1990). *Proceedings of European Conference,* Brussels.

Danish Working Environment Service. (1994). *Assessment of safety and health conditions at the workplace.* Copenhagen: Danish Working Environment Service.

Danish Working Environment Service. (1994). *Workplace assessment.* Copenhagen: Danish Working Environment Service.

Department for Professional Employees DPE, AFL CIO. (1986). *Here comes tomorrow. Technological change and professional, technical and office equipment.* Notes from the conference of Department for Professional Employees, Teaneck, NJ, April 11–12.

Dieckman, H. (1992). German safety rules. *Datamation*, 22.

DIN. (1985b). *Application software; principles of testing.* Deutsch Institut fur Normung e.v., DIN 66285.

DIN. (1980). *Body dimensions of people.* Deutsch Institut fur Normung e.v., DIN 33402, Part 2.

DIN. (1980–1984). *Display Work Places.* Deutsch Institut fur Normung e.v., DIN 66234, Parts 1-10.

DIN. (1985a). *Display work stations—Principles of dialogue design.* Deutsch Institut fur Normung e.v., DIN 66234, Part 8.

Dixon, J.L. (1989). The Canadian Standards Association and the new CSA national standard on office ergonomics (CAN/CSA-Z412-M89). *Proceedings of the Human Factors Association of Canada 22nd Annual Conference,* 117–124.

Dreyfuss, H. (1973). *Humanscale.* Henry Dreyfuss Associates.

Dumas, J.S. (1988). *Designing User Interfaces for Software.* Englewood Cliffs, NJ: Prentice Hall.

Dzida, W. (1989). The development of ergonomic standards. *SIGCHI Bulletin,* 29(3), 35–43.

ECMA. (1985). *Ergonomics—Recommendations for monochromatic visual display devices.* European Computer Manufactures Association Standard, ECMA-10.

ECMA. (1984). *Ergonomics—recommendations for VDU workplaces.* European Computer Manufactures Association, Technical Report TR/22.

Edström, R. (1986). Regulating VDU workplaces. *International Scientific Conference, Work with Display Units,* 987–988.

Engle, S.E., and Granda, R.E. (1975). *Guidelines for man / display interfaces,* IBM TR 00.2720. NY: Poughkeepsie Laboratory.

European Community. (1990). Council directive, Minimum safety and health requirements for work with display screen equipment. *Official Journal of the European Communities,* No. L/156/14.

FIET. (1985). *International trade union guidelines on visual dis-*

play units. Geneva: International Federation of Commercial, Clerical, Professional, and Technical Employees.

Foote-Lennox, T. (1986). Ergonomic guidelines for computerized user interfaces. *Computer Standards and Interfaces,* 5, 195–199.

FPTA. (1980). *Safety regulations for display workplaces in the office sector.* Injuries Insurance Institute for Banks, Insurances, the Free Professionals, and Specialized Undertakings, ZH1/618. Central Agency for Accident Prevention and Industrial Health Care.

FPTA. (1976). *Safety regulations for office workplaces.* Injuries Insurance Institute for Banks, Insurances, the Free Professionals, and Specialized Undertakings, ZH1/535. Germany: Central Agency for Accident Prevention and Industrial Health Care.

Friedman, A., Greenbaum, J., and Jacob, M. (1984). The challenges of users and unions. *Datamation,* 93.

Fuller, J. (1990). *The workplace workbook: An illustrated guide to job accommodation and assistive technology.* Washington, DC: The Dole Foundation.

Gorrell, E.L. (1980). *A human engineering specification for legibility of alphanumeric symbology on video displays* revised. Canadian Defense and Civil Institute of Environmental Medicine DCIEM. Tech. Rep. No. 80-R-R-26.

Granda, R.E. (1994). A methodology for industry conformance testing and compliance to international health and safety VDU standards. *Work with Display Units,* 2, A22–24.

Grandjean, E., and Vigliani, E. (Eds.). (1980). *Ergonomic aspects of visual display terminals.* London: Taylor and Francis.

Greeson, J.C. (1990). International standards challenge flat-panel displays. *Information Display,* 6, 7–8.

Grudin, J. (1988). The standards factor, ANSI technical advisory group X3Vi and user interface standards. *SIGCHI Bulletin,* 20(1), 16–19.

GSA. (1987). *Access to information technology by users with disabilities.* General Services Administration, U.S. Department of Education, Section 508.

GSA. (1991). *Managing information resources for accessibility.* U.S. General Services Administration, U.S. Department of Education.

Gueant, J.J., and Gibellieri, E. (1994). The contribution of the European trade unions and of the TUTB to the design of VDT-assisted workplaces. *Work with Display Units,* 2, A28–29.

Haider, M., Slezak, H., Hohller, H., Kundi, M., Schmid, H., Stidl, H.G., Thaler, A., and Winter, N. (1975). *Arbeitsbeanspruchung und Augenbleastung an Bildschirmgeraten.Vlg.des O.G.B., Automatisationsausschuss der Gewerkschaft der Privatangestellten* [*Work stress and visual stress from VDUs*], Wein.

Helander, M.G. (1981). *A critical review of human factors standards for visual display units.* Westlake Village, CA: Canyon Research Group.

Helander, M.G. (1984). Standards corner. *Human Factors Society Bulletin,* 27(11), 3.

Helander, M.G., Billingsley, P.A., and Schurick, J.M. (1984). An evaluation of human factors research on visual display terminals in the workplace. In Muckler, F.A. ed., *Human Factors Review 1984* (pp. 55–129). Santa Monica, CA, The Human Factors Society.

Helander, M.G., and Rupp, B.A. (1984). An overview of standards and guidelines for visual display terminals. *Applied Ergonomics,* 15(3), 185–195.

Hewlett-Packard. (1990). *Software usability design guidelines.* Palo Alto, CA: Author.

Hirsch, R.S. (1982). National standards for the design of visual display terminals. *Proceedings of the Human Factors Society,* 290–294.

Horton, W.K. (1990). *Designing and writing online documentation,* New York: John Wiley and Sons.

Human Factors Society. (1987). *HFS/HCI Standards Committee Draft Proposal, V 1.2.*

IBM. (1991). *Common user access advanced interface design reference.* Gary, NC: Author.

ILO. (1989). *Working with visual display units.* International Labor Office, Geneva, Occupational Safety and Health Series, No. 61.

ISO/CEN. (1991). *Agreement on technical cooperation between ISO and CEN* (Vienna Agreement). Announcement published by the International Organization for Standardization and European Committee for Standardization.

ISO. (1992). ISO 9241, *Ergonomic requirements for office work with visual display terminals, part 1: General introduction.* International Organization of Standardization.

Part 1, General introduction, 1992.
Part 2, Guidance on task requirements, 1992
Part 3, Visual display requirements, 1992.
Part 4, Keyboard requirements, 1994.
Part 5, Workstation layout and postural requirements, 1994.
Part 6, Environmental requirements, 1994.
Part 7, Display requirements with reflections, 1992.
Part 8, Requirements for displayed colours, 1994.
Part 9, Requirements for non-keyboard input devices, 1994.
Part 10, Dialogue principles, 1994.
Part 11, Usability statements, 1994.
Part 12, Presentation of information [coding and formatting], 1994.
Part 13, User guidance, 1994.
Part 14, Menu dialogues, 1994.
Part 15, Command dialogues, 1994.
Part 16, Direct manipulation dialogues
Part 17, Form filling dialogues, 1994.

Karat, J. (1986). A brief history of the Human Factors Society human-computer interaction standards committee. *Computer Systems Technical Group Bulletin,* 13(4), 10–18.

Koizumi, N. (1994). A new Japanese industrial standard (JIS) corresponding to Part 3 of ISO 9241 Visual Display Requirements. *Work with Display Units,* 2, A13–14.

Koizumi, N. (1994). Reconsiderations of the VDT ergonomics standards from the technological, scientific, and Asian cultural viewpoints, but not the political one. *Asian Symposium on Information Display.*

Macintosh, D.J. (1994). U.S. VDU public policy measures, 1993, *Work with display units,* 2, A20–21.

Martin, R. (1991). Testing to meet the Swedish VDT standards. *Compliance Engineering,* 56–63.

Mayhew, D.J. (1992). *Principles and guidelines in software user interface design,* Englewood Cliffs, NJ: Prentice Hall.

Melbourne and Metropolitan Board of Works. (1984). *Guidelines for the introduction and use of screen based equipment,* 17, *Equipment software criteria.*

MIL-STD-1472D. (1994). *Military standard: Human engineering design criteria for military systems, equipment and facilities.* Washington, DC: U.S. Government Printing Office, Department of Defense.

Mountfield, B.A. (1992). Health and safety implications of European Community 1002 (EC 92). *Am. Ind. Hyg. Assoc.,* 53, 736–741.

MPR. (1988). *Guide to the evaluation of reports on the testing of visual display terminals.* Sweden: National Council for Metrology and Testing.

MPR. (1987). *Testing visual display units—Emission quantities.* Sweden: National Council for Metrology and Testing.

MPR. (1987). *Testing Visual Display Units.* Sweden: National Council for Metrology and Testing.

National Research Council. (1983). *Video displays, work and vision.* Washington, DC: National Academy Press.

National Swedish Board of Occupational Safety and Health. (1983). *Work postures and working movements,* Sweden: Ordinance AFS, 6.

NBOSH. (1984). *Ordinances of the National Board of Occupational Safety and Health, with Instructions on Work with Data Processing Units.* National Bureau of Occupational Safety and Health, No. 632,2 TI745–83.

NBOSH. (1981). *Potential health hazards of video display terminals.* National Bureau of Occupational Safety and Health, Publication No. 81–129.

NBOSH. (1979). *Reading of display screens.* National Bureau of Occupational Safety and Health, Directive No. 136.

Nishiyama, K., and Watanabe, S. (1986). Recommendations of Japan Association of Industrial Health on VDT work and occupational health. *International Scientific Conference, Work with Display Units,* 989–992.

OSHA. (1993). *Ergonomic hazards—prevention of cumulative trauma disorders,* Section 5110, Title 8, California Code of Regulations, Division 1, Chapter 4, Subchapter 7, Group 15, Occupational Hazards and Ergonomic Hazards.

Pedersen, V. (1994). *The Working Environment Act,* SAS Technical paper. Scandinavian Airlines System.

Pew, R.W., and Rollins, A.M. (1975). *Dialog specification procedures.* Bolt, Beranek, and Newman, Report No. 3129.

Powell, J.E. (1990). *Designing user interfaces.* San Marcos, CA: Microtrend Books.

Price, H.E. (1994). A western view of ergonomics in Japan. *HFES Bulletin,* 37(6), 1–2.

Reading, V.M. ed. (1978). *Visual aspects and ergonomics of visual display units.* London: Department of Visual Science, Institute of Ophthalmology, University of London.

Reed, P. (1988). A brief summary of the background and current status of the HFS human-computer interaction standards committee. *SIGCHI Bulletin,* 19(3), 25–27.

Reed, P. (1986). Human factors standards for the software user interface, Genesis and evolution of the HFS human-computer interface committee. *Proceedings of the Human Factors Society,* 2, 1409–1411.

Reed, P., and Williams, J. (1988). Report on the meeting of the international standards organization's working groups on software ergonomics and man-machine dialog. *SIGCHI Bulletin,* 20(2), 13–15.

Rupp, B.A. (1980). *Human factors of workstations with display terminals.* IBM Corporation Report G320-6102-1, San Jose, CA: IBM.

Sawdon, D. (1994). Whatever happened to MPR3? *Work with Display Units,* 1, E12–13.

Schell, D.A. (1986). Usability testing of screen design: Beyond standards, principles, and guidelines. *Proceedings of the Human Factors Society,* 2:1212–1215.

Schneiderman, B. (1987). *Designing the user interface.* Reading, MA: Addison-Wesley Publishing Company.

Schurick, J.M., Helander, M.G., and Billingsley, A. (1982). Critique of methods employed in human factors research on VDTs. *Proceedings of the Human Factors Society 26th Annual Meeting,* 295–299.

Smith, S.L. (1986). Standards versus guidelines for designing user interface software. *Behavior and Information Technology,* 5(1): 47–61.

Smith, S.L., and Mosier, J. N. (1986). *Guidelines for Designing User Interface Software.* Mitre Technical Report ESD-TR-86-278. Springfield, VA: National Technical Information Service.

Smith, W.J. (1981). Research on the impact of computer displays on the human visual system. *International Conference on Visual Psychophysics,* 2–3, 40–42.

Smith, W.J. (1983). *Public health and video display terminals.* Paper presented at the American Public Health Association Conference.

Smith, W.J. (1984). Who dictates your terminal design. *SIGCHI Bulletin,* 16(1), 6–11.

Smith, W.J. (1985). Sociopolitical factors are also important, spotlight on ergonomics. *U.S. Government Computer News,* 4(3), 35–37.

Smith, W.J. (1988). Color in Screen Display, A draft international standard. *SIGCHI Special Interest Group Computer and Human Interaction Bulletin.*

Smith, W.J. (1988). Standardizing color for computer screens, *Proceedings of the Annual Meeting of the Human Factors Society,* 1381–1385.

Smith, W.J. (1988). Update on VDT comfort and health issues. *Supergroup Magazine.*

Smith, W.J. (1989). Standardization of colors for computer display images. *Displays,* 185–191.

Smith, W.J., and Cronin, D.T. (1993). Ergonomic test of the Kinesis keyboard. *Proceedings of the Human Factors Society,* 1, 318–322.

Snyder, H.L. (1983). Visual ergonomics and VDT standards. *Digital Design,* 24–30.

Snyder, K.M. (1991). *A Guide to Software Usability,* White Plains, NY: IBM.

Swedish National Board of Occupational Safety and Health. (1979). *Reading of Display Screens.* Health Directive 136. Stockholm.

Stewart, T. (1991). *Directory of HCI Standards.* London: Systems Concepts.

Stewart, T. (1994). The CEN/ISO experience. *Work with Display Units,* 2, A1–2.

Stewart, T. (1989). VDU ergonomic regulations—A European perspective, (eds.) L. Berlinguet and D. Berthelette, *Work with Display Units* (pp. 473–480). 89, North-Holland: Elsevier Science Publishers, B.V.

Strambler, J.H. (1993). *The dictionary for human factors,* London: CRC Press.

Swedish Trade Union Confederation. (1983). *Health hazards in the working environment.* Sweden.

TCO. (1985). *Trade unions and the working environment.* The Central Organization of Salaried Employees in Sweden.

Thorell, L.G., and Smith, W.J. (1990). *Using computer color effectively.* Englewood Cliffs, NJ: Prentice Hall.

Treasury Board of Canada. (1983). *Ergonomic guidelines for microelectronic installations.* Treasury Board of Canada T.B. No. 787875.

Treasury Board of Canada. (1983). *Video display terminals—A guide for managers and supervisors.* Treasury Board of Canada Bulletin 16–83.

TUC. (1978). *Interim guidelines on operation of video display units.* London: Society of Graphical and Allied Trades, Trade Union Congress.

Turner, J.A. and Karasek, R.A. (1984). Software ergonomics, Effects of computer application design parameters on operator task performance. *Ergonomics,* 27(6), 663–690.

Vigone, M. (1994). CEN and standardization in the field of VDTs. *Work with Display Units,* 2, A7–12.

Voskamp, P., and Weeda, C.E. (1994). Implementation of the EEC Directive 90/270 in the Netherlands. *Work with Display Units,* 2, A15–16.

Webb, R.D.G., and McCarthy, J. (1984). Canadian standards association guidelines on ergonomics in office systems. *Proceedings of 1984 International Conference on Occupational Ergonomics,* 105–107.

World Health Organization. No. 99, *Visual display terminals and workers health.* Geneva: WHO Offset Publications.

Index

Access to Information Technology by Users with Disabilities, 138, 294
accident prevention rules, 97 (*See* Germany)
ACM (*See* Association for Computer Machinery)
ACTU-VTHC (*See* Australia Council of Trade Unions and Victorian Trades Hall Council)
ADA (*See Americans with Disabilities Act*)
adjustability requirements for
  chairs, 176, 178–179
  displays, 131, 171–173
  keyboards, 174
  workstations, 176–180
advisory committee of CAL-OHSA, 143–145
AFL-CIO, 55
AFNOR (*See* France)
AFSC DH (*See* Air Force Human Factors Engineering Series in United Military standards)
Agreement on the European Economic Area, 86, 291
air
  currents, 279
  purification, 279–280
  requirements, 182
  ventilation, 279
alternative compliance, 115
  methods, 83
  test methods, 80
ambient conditions
  lighting, 276, 181
  temperature, 182, 279–280
American Academy of Orthopedic Surgeons, 142
American National Standard for Human Factors Engineering of Visual Display Terminal Workstations, 114–124 (*See also ANSI/HFS 100*)
American National Standards Institute, 1, 3, 7, 8, 11, 13, 24–34, 46, 114, 148, 290
  activities, 26–27
  and ISO, 32–33
  and the U.S. government, 31–33
  committee intent, 114
  committees, 30
  consensus, 28
  harmonization with ISO HFES 100, 3, 7–8, 33, 112–114 122
    applications, 115
  committee representation, 117
  conformance, 119
  member composition, 114
  revision, 117
HFES 200 Human Computer Interaction standard, 122, 124
  international standards approval, 29
  member participation, 27–29
  membership, 28
  standards
    applications, 114
    approval, 31
    developers, 29–31
    submission, 30
    subcommittees, 30
  VDT ergonomic standard, 8, 51,
  X3V1.9, 127 (*See also* ANSI Z-365), 131, 148
American Society for Testing and Materials, 290
Americans with Disabilities Act, 133, 135, 274
  applications, 136
analytical evaluations, 266
angle of keyboard tilt, 174, 274
animated on-line help, 280
APEX (*See* Australian Council of Trade Unions and Victorian Trades Hall Council)
Apple, 124, 272, 274
Apple Macintosh, 6
arm
  load, 240
  pronation, 274
arranging components, 278
Association for Computer Machinery, 24, 25
Association of German Engineers, 95
Association of Professional, Executive, Clerical and Computer Staff, 54
Australia, 10, 37, 151–156
  Codes of Practice, 107
  Council of Trade Unions and Victorian Trades Hall Council, 54
  VDT standard, 156
Austria, 10, 39

background lightness and visibility, 62
Basic List of Definitions of Human Body Dimensions for Technical Design, 292
behavioral principles, 121

Belgium, 10, 39
Berufagenossen-schaften, 97
bill from U.S. House of Representatives, 147
biomechanic
  assessment, 258
  and physiological problems, 43
  capabilities and limitations, 68
  effort, 239, 242, 258
  instrumentation, 241
  limits, 239
  load of during keyboarding, 241
  measurement methods, 120
  problems, 38
blood circulation, 241
Board for Measurement and Testing, 102 (*See also* Sweden)
Board of Standards Review, 290
Bolt, Beranek and Newman, 47, 121
branches of the U.S. military service, 113
brightness, 296
  and color contrast, 275
British Colombia, 130
  checklist, 131
  regulation, 131
  workers' compensation claims, 130
British Health and Safety Executive, 37
British Standards Institution, 25, 49, 57, 109, 290
BS 7179, 109
BSI (*See* British Standards Institution)
budgeting for ergonomic functions, 271
button requirements, 296
  location, 274
  size, 274

cable management, 278–279
CAD/CAM, 67
California
  CAL OHSA, 131, 145
    committee findings and recommendations, 144
  CTD Bill, 146
  VDT Study, 143
  Code of Regulations: Occupational Noise and Ergonomic Hazards, Ergonomic Hazards—Division of Occupational Safety and Health, 145
  Occupational Health and Safety Administration,

California (cont.)
   143 (*See also* California CAL OSHA)
   Occupational Safety and Health Standards Board, 146
   Prevention of Cumulative Trauma Disorders (Ergonomics), 146
   safety standards, 143
   VDT legislation, 52
Canada, 10, 37, 111
   CAN/CSA-Z412–M89, 129
   harmonization, 131
   province regulation, 130–131
   standards, 128
      association, 291
      committee membership, 130
cataracts, 43
cathode ray tube displays, 62, 85
CBEMA, 33, 48 (*See also* Computer Business and Equipment Manufacturers Association)
CCITT (*See* International Telegraph and Telephone Consultative Committee)
CD (*See* committee draft standard)
CE approval mark/ symbol, 88
CEI (*See* Italy)
CEN (*See* Center of European Normalization/Standardization)
CENELEC (*See* European Committee for Electrotechnical Standardization)
Center for European Normalization/ Standardization (CEN), 20, 86, 21, 25, 49, 50, 86, 94, 110
   adoption of ISO 9241, 89
   certification, 88
   standard (CEN 29241), 94
   standards creation, 95
   TC 122, 95
   testing, 88
Central Organization of Salaried Employees (TCO), 54, 292 (*See also* Sweden)
certification
   agencies and symbols, 105
   certification labels, 98
   certification methods, 288
   certification test agencies, 98
chairs, 36, 43, 131
   requirements, 478–479
character
   height, 48
   illumination, 42
   readability, 276
   size, 276
      for flat panel displays, 275
check box, 306
check lists, 163–237
Chernobyl, 38

choice list, 306
CIE (*See* Commission of Illuminating Engineering)
civilian standards, 111 (*See also* standard types)
cluster workstations, 280
Codes of Practice: *Management of Health and Safety at Work* and *Display Screen Equipment Work*, 107 (*See also* Australia)
coding, 71
Coding of information (*See* German standards)
cognitive
   components of color interpretation, 67
   functioning, 279
   load, 241
   processing, 240
color, 141
   character recognition test, 263
   detection, 296
   difference calculation, 296
   discrimination, 296
   display standards, 22
   displays, 118
   identification, 296
   interpretation, 296
   matching test, 264
   thin fonts, 43
   vision deficiencies, 275
Colorimetric Illuminants, 293
Colorimetry, 293
Colors of Indicator Lights and Push-buttons, 294
comfort, 63, 83, 242
   assessment, 261
command button, 306
commands, 281
   dialogues, 73, 202
   languages, 69, 121
   representation specifications, 73
Commission Internationale de l'Eclairage (*See* Commission of Illuminating Engineering)
Commission of Illuminating Engineering, 7–8, 11–12, 18–19, 290
Commission of the European Communities, 50, 290
committee draft standard, 12, 13, 76
common market, 85
comparative questionnaire/ rating scale, 261, 263
competitive strategy, 270
complaints from VDT users, 144
compliance, 80
   assessments in ISO 9241, 81
   components of ISO 9241, 170
   implications for industry, 94
   measurements, 115

   of a product, 238
   specifications in Parts 1–17, 80
   with an CEN standard, 88
   with ISO 9000, 287
   with ISO 9241, 287
composition
   of EFTA, 85
   of ISO 9241, 58
   of the EU, 85
computer
   aided design, 42, 67
   applications of GSA 508, 139
   industry, 129
   interface, 232
   system, 247
Computer Business Equipment Manufacturers Association, 33, 48, 290
Computer Human Interaction, 290
   standard, 125
Confederation of Professional Employees, 103
conformity of European countries to VDT Directive, 94–95
Considerations in the Design of Computers and Operating Systems to Increase their Accessibility to Persons with Disabilities, 141
consumer product ergonomic standards, 288
contrast, 42, 276
   between characters and background, 62
   ratio, 296
Control of Work-Related Cumulative Trauma Disorders, 148, 292
control panels, 274, 276
convergence, 297
corporate
   objectives, 282
   strategies, 271
correlation
   between these disorders and display use, 276
   of comfort with performance, 241
cost
   and function of computer software, 281
   business and government, 130
   concerns, 44
   low profile, detached keyboard, 44
   testing, 283
Council for Mutual Economic Assistance, 290
Court of Appeals, 146
CRT Display and Keyboard Units for Business Use (JIS 6041), 159, 295
CSA (*See* Canadian standards)

CTD (*See* cumulative trauma disorders)
cumulative trauma disorders, 52, 130, 142, 145, 240
  regulation, 131
  United States, 143

de facto ergonomic standards (*See* ergonomic standard types)
default
  button, 306
  color palettes, 275
  fonts, 277
defeat of U.S. proposed VDT legislation, 50
defective color vision, 297
Denmark, 10, 106
  conformance penalties, 106
  Work Environmental Service, 106
Department of Agriculture (*See* United States)
Department for Professional Employees, 55
Department of Defense, 122, 291 (*See also* United States)
  standard, 122
Department of Education, 138 (*See also* United States)
Department of Transportation, 291 (*See also* United States)
depth-of-field, 297
design
  computer workstations, 111
  for the disabled, 274
  goal, 240
  test measures, 239
  validation, 288
  viewing distance, 297
  weaknesses, 283
  working position, 297
designers, 93, 112
desk and chair requirements, 216
detached and low profile keyboards (*See* keyboards)
Deutsches Institut fur Normung (DIN), 7, 13, 24, 25, 95, 98, 291 (*See also* Germany standards)
  66234, 36, 37, 98
  IEC 380/VDE 0806, 96
  IEC 65/VDE 0806, 96
  safety standards, 95
development and marketing, 281
dialog
  box, 306
  interaction standards, 18
  Principles, 69, 188
difference
  between North America and Europe standards, 132
  between technical and ergonomic standards, 44
  in philosophy and content between the U.S. and Canadian ergonomics standards, 129
Direct manipulation dialogues, 73, 204
directive, 10, 40, 293 (*See also* requirements)
  applications, 92
DIS (*See* draft international standard)
disabled persons standards, 21
discomfort, 239, 279
disorders, 239
displays, 103, 118, 240
  features specified and tested by MPR, 103
  frame layout, 121
  requirement topics and measurements, 115
  requirements, 48, 61, 115, 171, 211
  requirements with reflections, 66, 183
  resolution, 43
  screen, 109, 230
  standards, 21–22, 33
  usability test, 261
Display Screen Directive, 90 (*See also* EU)
Display Screen Equipment Work, 294
Display Specifications Procedures (*See* software guidelines)
Display standards, 22
Display Workplaces in the Office Sector, 295
distribution of non-complying products, 97
District Public Works (*See* Australia)
document holders, 36
documentation compliance, 141
documented evidence, 266
documents, 36, 42
DOD, 25, 127 (*See also* United States military standards)
DOD-STD-2167, 113 (*See also* United States military standards)
DOE, 127 (*See also* United States standards)
DOT, 25, 127 (*See also* United States standards)
draft
  addendum, 295
  amendment, 295
  international standard, 12, 13, 76, 295
  proposal, 295
  technical report, 295
dragging, 297
  task, 256
  test, 256

ECC, 110
ECMA (*See* European Computer Manufacturers Association)
edge
  sharpness, 43
  visibility, 276
EEA, 86 (*See also* European Economic Area)
EEC, 85 (*See also* European Commission)
EFTA (*See* European Free Trade Association)
electric and magnetic fields standards, 22
electrical equipment, 96
electromagnetic fields, 65
emission protection law, 96
empirical evaluations, 266
employers, 10, 93, 134, 271
  practice toward VDT workers, 53
  responsibility to ADA, 137
  regulations, 131
  requirements, 91
  obligations, 228
employment of ergonomists, 271
EN(s) (*See* European norm(s))
environment, 35, 231
  characteristics and measurements, 117
  Environmental Labeling of Displays, 105
  requirements, 65, 92, 181, 217
  standards, 6, 16, 33
Equipment, 229
  arrangement, 36
Equipment Safety Law, 96, 97, 295 (*See also* Germany)
Ergonomic Aspects of Visual Display Terminal, 37
ergonomics, 3
  associations, 22–24
  certification, 288
  conferences, 38
  design process, 282
Ergonomics (*cont.*)
  explosion, 269
  guidelines, 121
  in the U.S., 41
  requirements for managers, 128
  research, 43, 132
  standard expansion, 37
  standard developers, 5
Ergonomics of design and use of visual display terminals (VDTs) in offices, 292
Ergonomic Principles Related to Mental Workload—General Terms and Definitions, 158
Ergonomic Requirements for Office Work with Visual Display Terminals (VDTs)—Visual Display (*See* ISO 9241)

Ergonomics—Anthropometric and Biomechanic Measurement, 158, 295
Ericksson, 272
error and help messages, 219
error prevention, 121
errors message specifications, 47
ESPRIT (*See* Europe Strategic Program of Research Development in Information Technology)
ETSI (*See* European Telecommunications Standards Institute)
EU (*See* European Union)
Europe, 10, 4636, 56, 131, 151
  accreditation, 88
  competitive edge, 110
  Commission, 88
  Committee for Electrotechnical Standardization, 20, 50, 86, 102, 290
  Committee for Standardization, 290
  Computer Manufacturers Association, 21, 22, 291
  Council, 9, 89
  Council of Ministers, 90
  country VDT legislation, 110
  Court of Justice, 88
  directive, 20, 89, 110
    application, 90, 93
    definition, 89
    requirements, 227
    display screen, 40, 132, 271
  Economic Area, 86, 94
  Economic Community, 84, 85
  ergonomics, 5
  Free Trade Association, 21, 84–85, 94, 110, 291
  GNP, 84
  harmonization, 95
  information technology industry, 110
  norms, 94, 95, 295
    EN 29241, 95
  organizations, 84
  research, 110
  solution to inconsistent standards, 49
  standard organizations voting on ISO 159, 87
  standards agencies, 20–22, 86
  social charter, 89
  Strategic Program of Research Development in Information Technology (ESPRIT), 110, 291
  Telecommunications Standards Institute, 21–22, 291
  Union, 8–9, 10, 21, 50, 84–85, 94, 110, 284, 291
    directive implementation, 84, 106

Examples of
  balanced test subject assignment, 249
  CRT specifications in JIS 6041, 159
  compliance methods, 82
  IBM action message guideline, 126
  Macintosh error message guideline, 125
  Mitre guideline format, 123
  national and international ergonomic standards, 7
  national and international VDT standards, 8
  physical impairments, 134
  required keyboard features, 63
  requirements and recommendations, 78
  standards agencies and VDT ergonomic standards, 25
  test stimuli matrix, 264
executive management support for ergonomics in the life cycle, 282
expansion of European standards and legislation, 287
expert systems, 288
extension (hand), 274
eye
  dryness, 279
  exams, 50, 108
  stress, 54
  tests, 106

face rashes, 54
factors
  in liability cases, 152
  influence ergonomic standards, 287
federal agencies, 138
federal laws/ regulations, 134, 270
Federal Minister of Labor and Social Affairs (*See* Germany)
Federal Ministry of Labor, 97
Federal Occupation Safety and Health Administration, 147
feedback, 219, 298
FIET, 53
finger abduction, 298
Finland, 10, 39
first ergonomic ordinance, 96
fiscal barriers, 85
flat panel displays, 77, 85, 118, 275
  standards, 22
flicker, 240
  standard (*See* Japan)
focus groups, 283
fonts, 277
footrests, 36
form based entries, 69
Form filling dialogues, 74, 208

format specifications, 47
formatting, 71
Fourier transforms, 120
Framework Directive (Article 16), 90
France, 10, 37, 39
  AFNOR, 390
  standards agency, 25, 390
free hand input, 298
  input test, 257
  symbol entry task, 258
functionality, 278
funding of ergonomic research, 288
furniture, 118
  features and measurements, 116
  standards, 33, 287

General Services Administration, 25, 52, 138, 291 (*See also* U.S.)
geometric design of characters, 98
German (Deutsch) Institute of Normalization, 7 (*See also* Germany)
Germany, 10, 35, 39, 42, 44, 56, 95, 114
  accident prevention rules, 97
  belief of standard basis, 46
  electrical standard, 96
  Electrotechnical Committee, 95
  ergonomic ordinance, 96
  implosion standard, 96
  Industrial Injuries Institute, 53
  Institute for Standardization, 95
  laws and regulations, 96
  legal aspects of regulations, 97
  safety (GS) test mark, 98
  safety regulations, 96
  standards
    DIN 66234
      Coding of information, 99
      Grouping and formatting of data, 99
  standards-development agencies, 95
glare, 42
glasses for VDT viewing, 106
GNP and population of EU and EFTA, 85
goal
  EU directive 90/270, 90
  ANSI VDT standard, 114
government
  agencies, 11
  agency compliance, 53
  organizations, 129
graphic
  design, 277
  digitizer, 299
  tablet, 299
  in on-line help, 281

processing applications, 118
grasp and park (homing) test, 257
gray and black background screens, 43
Great Britain, 39 (*See also* United Kingdom)
GS mark combined with VDE test agency, 98
GS safety mark, 98
GSA (*See* General Services Administration)
Guidance Note for the Prevention of Occupational Overuse Syndrome in Keyboard Employment, 156
Guidance on specifying and measuring usability, 69
guidelines, 1, 8–9
  GSA 508, 139, 141
  hearing disabled, 235
  microelectronic technology for federal government users, 128
  product user interfaces for disabled individuals, 164
  user interfaces for individuals with disabilities, 233
    moderately disabled, 233
    severely disabled, 234
  users with seizure disorders, 235
  visually impaired, 234
Guidelines for Designing User Interface Software (*See* Mitre Corporation)
Guidelines to Occupational Health in VDT Operation (*See* Japan)
Guidelines for the Introduction and Use of Screen Based Equipment (*See* Australia)
Guidelines for Man/Display Interfaces (*See* IBM)

hand
  angles, 240
  extension, 298
  flexion, 299
hardware
  and software requirements, 111
  issues, 42
  requirements, 91, 163, 211
harmonization of standards, 17–18, 21
HCI report created by the ANSI committee, 126
HCI standards (*See* software standards and the United States)
health agencies, 11
health and safety, 35, 58
  requirements for VDTs, 90

directive, 90
Health and Safety (Display Screen Equipment) Regulations, 108 (*See also* United Kingdom)
Health and Safety Commission, 107, 291
Health and Safety of Work Act, 107 (*See also* United Kingdom)
health
  effects of VDTs, 51
  hazards of electromagnetic emissions, 102
  issues, 287
  organizations, 102 (*See also* WHO)
  requirements, 142
hearings on VDT health and safety, 50
heat, 231
heat and air circulation, 279
Hewlett-Packard, 122, 282
HFES (*See* Human Factors and Ergonomics Society)
HFS (*See* Human Factors Society)
historical differences between ergonomics in Europe and the U.S., 46
House of Representatives (U.S.), 147
Human Computer Interaction committee (*See* United States)
Human Engineering Design Criteria for Military Systems, Equipment and Facilities, 112
Human Engineering Guidelines for Management Information Systems, 121
Human Engineering Requirements for Military Systems Equipment and Facilities, 113
Human Factors and Ergonomics Society, 5, 46, 291
human factors engineering, 5
  hiring, 272
  society, 23–24
Human Factors Engineering of Visual Display Terminal Workstations, 33, 292 (*See also* ANSI)
Human Factors Guidelines and Preferred Practices for the Design of Medical Devices, 122
Human Factors Society, 124, 291
Human Sensory Measurement Applications Technology project, 161
Human Technology Project, 161 (*See also* Japan)
humidity, 232, 279

IBM, 47, 121, 274
Guidelines for Man/Display Interfaces, 121
ICL, 272
icons, 277, 306
Graphical Symbols for Use on Equipment—Index and Symbols, 294
identification and interpretation, 277
ideograph characters, 159
IEA (*See* International Ergonomics Association)
IEC (*See* International Electrical Commission)
IEEE (*See* Institute of Electrical and Electronic Engineers)
illumination, 36
ILO (*See* International Labor Organization)
impact
  of ergonomics, 269
  of wrist rests, 240
implications of ISO 9000 for design, 286
importance of usability features, 271
importers, 95
incidence of skin problems, 280
increased interest in ergonomic standards, 275
independent questionnaire/rating scale, 261–262
industry
  agencies, 11
  conclusions, 144
  inspection boards, 97
  software guidelines and standards, 111
information presentation, 191
Information Technology Access by the Disabled, 138 (*See also* United States)
Information Technology Industry Council, 33–34, 46, 291 (*See also* United States)
initial reaction of industry to ergonomic standards, 271
input
  characters, 251
  data, 252
input devices, 113, 299
  compliance, 139
  requirements, 214
  keyboards, 63–64, 173–176
  non-keyboard, 68–69, 185–187
  strategies, 272
inspection of the equipment, 153
Institute of Electrical and Electronics Engineers, 25, 127, 291

Instructions about Working with VDUs, 106
integrate ergonomics into product development and marketing, 282
intention of ISO 9241, 58
interactive tutorial, 280
internal guidelines, 121
International Council of Societies of Industrial Design, 291
International Electrical Commission, 12, 17, 50, 95, 291
International Ergonomics Association, 22, 291
 federated societies objectives
International Federation of Commercial Clerical and Technical Employees, 291
International Labor Organization, 20, 23, 291
International Lighting Vocabulary, 293
International Organization for Standardization, 1, 7–8, 11–19, 21–23, 26–27, 29, 32–34, 49–50, 95, 291
 objective, 12
 scope, 12
 development process, 12–14
 technical committee, 49
 ISO 9000, 105, 284
  auditor, 284
  audits, 286
  certification, 284
  compliance, 285
  parts, 285
  requirements, 284
 ISO 9995 (*See* keyboard standards)
 ISO 9241, 46, 57–83, 86, 95, 99, 109, 110, 115, 122, 169–211, 238
  application domains of ISO 9241, 75
  checklists, 169–211
  organization by subject, 59
  parts, 60, 76, 169–211
  focus, 60
  status, 76
  revision, 17
  test protocols, 265
  Parts 16, 47, 127
 TC 159, 13, 14
  objectives, 14–17
  organization, 16
  subcommittees, 15
  working groups, 15
  participants, 308
international spread of VDT ergonomic issues, 39
international standards, 12, 13, 76, 295
International Telegraph and Telephone Consultative Committee, 19, 290
internationalization of software specifications, 47
intuitive user interfaces, 280
ISs (*See* international standards)
Italy, 37, 38, 39, 107
 Italian Electrotechnical Committee (CEI), 107
 first legislative decree, 629/94, 107
ITI (*See* Information Technology Industry Council)

Japan, 10, 37, 157, 161, 278
 adaption to ISO 9241, 158
 companies, 282
 ergonomic research, 161
 Ergonomics Research Society, 158
 ergonomic standard, 158
 flicker standard, 240
 guideline topics and requirements, 158
 Industrial Safety and Health Association, 157
 Institute of Standardization (JIS), 25, 291
 standard on CRTs and key boards, 159
 Institute of Standardization, 7
 Ministry of International Trade and Industry, 292
 standards, 13, 24
 JIS 6041
 JIS Z 8500
JERS (*See* Japan Ergonomics Research Society)
joint technical committee for ISO, 17, 18, 291
JTC (*See* joint technical committee)

key
 force, 240
keyboard, 6, 16, 18, 33, 230, 36, 103, 109, 118
 features and measurements, 116
 layout
 regulation, 155
 requirements, 48, 63, 173, 213
 specifications and implications, 64
 test, 250, 253
 standards, 16, 18, 21, 33
keying
 errors, 240
 input, 220, 250
 performance analysis, 252
 tasks, 252
keystroke monitoring, 50
keyword, 306
knowledge and skills required, 120

label size and contrast, 276
labeling on most product control panels, 276
labor
 arbitration, 9
 demands, 287
 negotiation, 9, 288
 organization, 11, 103
 representatives, 144
 unions, 9, 53
  activities, 40
  guidelines, 54
lap top
 computers, 77
 systems, 275
lateral hand deviation, 299
laws (*See* requirement types)
learning, 281
legal action, 96
legibility of software bundled fonts, 278
legibility testing of screen characters, 263–265, 277
legislation, 269, 288
 by the EU, 89
 for disabled users, 133
 in the United States, 50
 of VDT design and use, 55
light
 conditions, 275
 low ambient levels, 42
lighting, 231
on-line trainers, 288
Lisbon Agreement (*See* Europe)
litigation, 269, 270, 288
lobbying against VDT legislation, 50
local standard (*See* standard types)
lost revenue, 270

machinery standards, 95
Macintosh guidelines, 124
major ergonomic events in the last decade, 270
major events in ergonomic standardization, 39
major issues of the EU Social Charter, 89
Management of Health and Safety at Work, 294
management responsibilities, 138
mandatory standards (*See* standard types)
mandatory requirements, 80
 (*See also* requirements types)
manipulation, 306
manufacturers, 10, 83, 93, 95, 109, 112–113, 134
marketing, 47, 281–283
marketing tool, 270
measurement, 119
 methods, 44, 100
 techniques, 120
Measuring Methods of Phosphor Persistence for CRT

330

Screens, 160, 295 (*See also* Japan)
mechanical vibrations, 65
medical and scientific instrumentation displays, 67
medical product standards, 288
meeting process requirements of ISO 9241, 284
members of the EU and EFTA, 86
mental impairment, 134
mental stress, 38
menus, 220
   bar, 307
   choices, 226
menu dialogues, 72, 198
   selection specifications, 47
   titles, 221
menu-based dialogue principles, 126
methods to evaluate ergonomic characteristics of VDTs, 102
Metropolitan Board of Works (*See* Australia)
mice requirements, 186, 274, 300
microfilm readers, 36
Microsoft, 272
Microsoft Windows, 6
Milan, 37
military standards
   1472 requirement applications, 111–112
   establishment requirements, 112
   Medical Device Guidelines, 122
   MIL STD 1472, 112
   MIL-H 46855, 111, 113
   MIL-HDBK-759, 113
   MIS Guidelines, 121
   Standardization Handbook: Human Factors Engineering Design for Army Material, 113
miniaturization, 278
Minimum Safety and Health Requirements for Work with Display Screen Equipment (90/270/EEC), 90, 93
Ministry of International Trade and Industry (MITI), 157 (*See also* Japan)
MITI Consortium, 157
Mitre Corporation, 122.
   guideline format, 124
   guidelines, 118, 122
modular/cluster workstations, 275, 279, 280
monitoring equipment, 247
Motif, 6, 128
MPR, 101
   1, 102
   2, 102
      conversion of *MPR 2* into a CEN standard, 102

1990: 8, 102
1990:10, 102
   development committee, 102
multi-directional tapping task, 256
multi-media, 280, 288
multinational regulations, 86–95, 270
multiple test measures, 241
multi-regional legislation, 132
muscle effort, 259
   measurement, 260
musculoskeletal, 40
   discomfort and disorders, 144, 274
   disorders, 38, 55, 152
   functions, 276

NAS (*See* National Academy of Science)
National Academy of Science report, 41
National Board of Occupational Health and Safety, 40 (*See also* Sweden)
national
   standard agencies, 24–25, 308
   laws, 52
   legislation, 93
   standards, 89, 287
National Institute for Standards and Technology, 24, 25, 127–128, 292
National Occupational Health and Safety Commission (*See* Australia)
National Safety Council, 148
   ANSI Z365 standard, 148–149
national safety ergonomic standards, 147
natural and artificial lighting, 65
navigation, 222
Navy Defense System Software Development, 113 (*See also* United States)
NEC, 272
negative polarity, 43, 141
nerve compression, 241
neutral hand/arm position, 300
neutral posture, 69
new additions to ergonomic computer standards, 77
New York Committee for Occupational Safety and Health, 54
NIOSH, 41
NIOSH standard, 7
NIST (*See* National Institute of Standards and Technology)
non-keyboard input devices, 118, 185–187
noise, 231
Nokia, 272
non-keyboard

input device tests, 253–258
   standards, 16, 68–69, 185–187
non-normative (*See* standard types)
normative standards (*See* standard types)
normative requirements and implications, 61
North America, 10
   standards, 111–132
Norway, 10, 106
   Working Environment Act, 106
noxious substances, 113
NYCOSH (*See* New York Committee for Occupational Safety and Health)

objectives
   in publishing the *Screen Checker*, 105
   ADA, 136
   ANSI VDT revision committee, 118
   revision committee, 117
   CEN, 94
observational analysis, 81, 266
Occupational Safety and Health Administration, 147–148
   assessment methods of OSHA standard, 149
Occupational Safety and Health Association, 292
Occupational Safety and Health Standards Board (*See* California)
Office Ergonomics standard, 129, 292 (*See also* Canada)
office illumination, 42, 181
older users visual problems, 276
one direction tapping task, 255
on-line help, 280
on-the-job training, 106
Open Software Foundation, 127–128
operating instructions, 96
optimal performance, 240
option groups, 222
option selection and activation, 223
ordinance for handicap use, 52
ordinances for VDTs (*See* U.S.)
OSF (*See* Open Software Foundation)
OSHA (*See* Occupational Safety and Health Administration)
output device compliance, 140–141
oxygen deprivation, 241

P1201 committee, 128
Pacific Rim, 10, 162
padded wrist rests, 240
palm rests, 43

palm/wrist rest, 300
papers, 38
parallax, 300
partial ergonomic analysis, 283
participation of the U.S. in ISO standards development, 114
perceptibility of characters on screens, 99
perceptual components of color, 67
performance, 242
performance (*cont.*)
   and comfort test measures, 241
   evaluators, 288
   safety and health based standard, 49
Philips, 272
photometric instruments, 120
physical barriers, 85
physical impairment, 134
physically disabled individuals, 134
physiological and psychological testing, 70
   functioning, 242
   processes, 241, 276
pointing tests, 253
pointing, selection/activation, 223
polarity, 277
population of TCO, 103
portable computer systems, 92
positive
   display imaging, 42
   polarity, 43
postural deviation from neutral, 143, 258
   measurement, 258, 260
   problems, 54
preferences, 83, 277
pregnancy, 38
   outcomes, 144
   risks, 54
Preliminary Release of European Norm, 295
presentation of information, 70
pretest protocol for display tests, 263
pre-testing, 249
prevalence of older ergonomic requirements, 44
Prevention and Management of Occupational Overuse Syndrome (*See* Australia)
primary goal of 1472, 112
Principles of Human Behavior Related to Safety, 113 (*See also* United States military standards)
Principles of man-machine dialogue design, 99
printers, 278
process strategies, 281
procurement agencies, 113
product
   compliance measurement, 41
   design standards, 6

development process, 271, 282
discriminator, 288
flexibility, 278
identification, 246
portability, 278
set-up and reporting, 246
size, 278
strategies, 272
weight, 278
psychological stress, 54
purchase factor, 288
purchase requests, 271

Quality Assurance Standard, 294
quantity of text of FPD, 275
questionnaires, 162, 261–263
QWERTY (*See* keyboards)

radial deviation, 274
radiation emission, 38, 54, 103, 113, 162, 231, 276
   limits, 102
rating scales, 64, 261–263
recall of products, 283
reflections and glare, 231
   requirements, 183–184
refresh rate, 240
regional standard (*See* standard types)
Regulation for Display Work Places in the Office Sector, 36
Prevention of Cumulative Trauma Disorders (Ergonomics) (*See* California)
Rehabilitation Act, 138
relationship between effort, comfort, and performance, 239
repetitive strain disorders/injuries, 142, 152
repetitive task standards (*See* types of standards)
report on health issues, 40
reproduction, 276
   disorders, 40
relationship between display use and visual and reproductive disorders, 276
requirements
   ADA, 164, 232
   CAL-OHSA bill, 146
   color, 67, 184–185
   equipment for the disabled, 133
   GSA 508, 141
   ISO 9241, 78
   keyboards, 173–176
   non-keyboard input devices, 68, 185
   OSHA standard, 149
   software 188–211
   VDT hardware and software user interfaces, 58
   work organization, vision, posture, and work environment, 131

workstations, 176–180
research, 287
   needed, 277
   on VDT health and safety issues, 55
Research Institute of Human Engineering for Quality of Life, 161 (*See also* Japan)
resolution, 276, 302
response times, 121
responsibilities of CEN, 94
rest breaks, 108, 131
risk of errors, 112
RSI, 302
   cost, 142
   epidemic, 151
   liability, 152
   regulation, 154

safety
   and design guidelines, 164
   labels, 97
   law requirements, 96
   requirements, 163
   standards, 6
Safety and Health Directive, 90
Safety of Information Technology Equipment, including Electrical Business Equipment, 294 (*See also* Australia)
Safety Regulations for Office Working Places, 36
San Francisco
   Board of Supervisors, 145
   VDT bill, 148
   VDT ordinance, 145
scope of ISO 9241, 49
Screen Based Workstations, 292 (*See also* Australia)
Screen Checker, 103 (*See also* Sweden)
Screen Facts and *Software Checker*, 105 (*See also* Sweden)
screen
   flicker, 42
   height, 43
scrolling, 218
seated postures, 118
selection criteria for input test(s), 254
semantic specifications, 241
SEMKO, 105
shifts of terminal work, 106
Siemens, 272
simultaneous measurement of variables, 239, 241–242
skin, 40
   problems, 279
software, 232
   checklist, 224
   dialogue, 241
   ergonomic standards, 44
   guidelines, 217
   issues, 44
   requirements, 92
   specifications, 47, 188–211

standards, 124
test methods, 267
user interface design, 280
user interface tests, 265
user requirements, 163
Software Ergonomics and Man-Machine Dialogue, 127
sound and noise, 65
Soviet Republics, former, 161, 162
space requirements, 36
Special Interest Group on Computer and Human Interaction, 292
special standards (*See* United States)
specifications
 for system response time, 47
 of equipment design, 41
Spectroradiometric measurement of light sources, 293
split keyboards, 273
standards, 1
 agencies, 11
 benefits, 2
 compliance, 239
 development, 40
 focus in the U.S. (*See* United States)
 history, 35–56
 icons, 277
 impact and future, 269–289
 new technologies, 288
 legislation in Europe, 84
 objectives, 2
 organizations, 11–34
 philosophies, 45
 types
  civilian, 114–132
  international, 2
  local, 2
  mandatory, 2–3
  military, 7,
  national, 2
  non-normative, 83
  normative, 83
  technical, 6
state and county
 legislation, 50, 145
 legislation proposed on VDT equipment, 51
statistical analyses, 120
stereopsis, 302
strategic planning, 295
 corporate, 271–272
 product features, 272–281
 process, 281–283
strategy of ESPRIT, 110
strength measures, 259
stress, 38, 40, 239
structure and syntax requirements, 73
stylus, 303
stylus and tablet, 303
subcommittee, 295
subject treatment, 248
supination, 303
survey of customers, 271
Sweden, 10, 35, 39, 101

Board for Measurement and Testing, 102
Central Organization of Salaried Employees, 53
conference, 38
conferences, 40
guideline
 order, 102
ordinance specification, 101
TCO (Swedish Central Organization of Salaried Employees), 54, 103
 guidelines, 162
 specifications for displays, 104
 specifications for key boards, 104
Union Requirements, 103
symbols and icons standards, 18
system
 description, 232, 266
 effectiveness, 70
 individualization, 219
 modification requirements, 113
 standards, 21
 rather than a component, approach, 288

tablet
 with overlay, 303–304
 with stylus, 304
tactile feedback, 304
TAG (*See* Technical Advisory Group)
task and job design standards, 16
task requirements, 59, 61, 171
TC (*See* ISO technical committee)
TC 122 WG 5, 95
TC 159 development activities, 14 (*See* ISO)
TCO (*See* Swedish Central Organization of Salaried Employees)
Technical Advisory Group, 295
technical
 barriers, 85
 committees, 12, 87, 295
 technical report, 296
 standard, 44
 support, 280
Technical University of Berlin, 36
telecommunication standards, 19
test and testing, 235
 administrator, 243
 conditions and procedures, 243
 description, 261
 forms, 244
 instruments, 82
 label for VDE standards agency, 97
 measures, 240
 methods, 102, 105

methods for radiation emissions, 102
procedure, 63
procedures, 248
room requirements, 245
screen image perception, 120
session schedule, 253
station, 245
stimuli, 262
subject criteria for display test, 263
subjects, 244
task selection, 253
technicians, 267
Test Method for Visual Display Units, 102, 295
the usability of icons, 277
Text and Office Systems Applications (*See* ISO JTC/SC18)
text processing, 115
thermal conditions, 65 (*See also* heat)
time-to-market, 283
topics in the 1988 and ANSI/HFS 100 standard revision, 119
touch sensitive screen, 304
tracing task, 257
Trade Cooperative Association, 292
trade unions, 102
Trades Union Congress, 292
training, 108, 112
 and skill of test administrator, 243
 in usability testing and data analysis, 283
transfer to other jobs, 51
Treasury Board of Canada, 128
treatment and prevention of RSIs, 142
Treaty of Rome, 85, 94
TUB (*See* Germany and *see also* Technical University of Berlin)
Turin conference, 38
TUV, 98
type of measures, 258

U.K. ergonomic regulations, 107
U.S. (*See* United States)
ulnar hand deviation, 274, 305
union
 activities, 53–55
 beliefs, 55
 concern regarding VDT user health, 54
 software standards, 55
United Kingdom, 10, 107
 Health and Safety Executive, 53
United Nations, 20
United States, 10, 35, 41, 42, 44, 52, 111, 162
 Army, 121
 civilian standards, 114
 computer industry, 42, 44

333

United States (cont.)
  Congress, 53, 138
  ergonomic standards bodies, 25
  federal regulations, 41
  mandate from U.S. Congress, 53
  laws for the disabled, 52
  military standards, 7, 112
    Air Force Human Factors Engineering Series, 13, 113
    Military Standardization Handbook: Human Factors Engineering Design for Arm Material, 113
    Navy Defense System Software Development, 113
    Principles of Human Behavior Related to Safety, 113–114
  state bills, 270
  Congress, 50
  Senate, 50
  standards, 111
  standards for special circumstances, 133–150
  standards agencies, 24
UNIX, 281
unpadded palm rests, 43
usability
  guidelines, 190
  testing, 238–268
  advantage, 267
  measures, 242
User's Handbook for Evaluating Visual Display Units, 102
user
  control, 218
  expectations, 218
  User Guidance, 71, 195
  interface, 307
  performance, 63, 83, 279
  performance measures, 81
  system interfaces and symbols standard (*See* ISO JTC/SC18/WG9)
users with special requirements, 288
User-system Interfaces and Symbols, 127

validation of designs, 283
VDE, 95, 98
VDI, 95
VDT
  characteristics, 100
  conferences in Europe, 38
  proposed legal requirements, 50
  tables, 36
  users with musculoskeletal problems, 152
  viewing glasses, 50
  Work Stations, 36
  workstation standard, 7
VDU Work Places standards, 22
Victorian Trades Hall Council, 54
Video Displays, Work and Vision (*See* National Academy of Science report)
Video System, 247
Vienna Agreement, 49, 95
viewing
  angle, 276
  distances, 276, 277
violation of Italian VDT law, 107
violation of the ADA, 137
virtual reality, 288
visibility of labeling, 274
vision, 40, 276
  low visual acuity, 43, 274
visual, 40
  adaptation, 305
  and cognitive aspects of icons, 277
  and neck discomfort, 43
  annoyance and complaints, 43
  comfort, 42
  display performance, 62
  display units, 36
  exams, 36
  issues, 41
  performance, 42, 240
  problems, 38
  stress, 42
  impaired, 141
Visual Display Terminals and Worker Health (*See* WHO)
Visual Display Unit Work Stations, 293

voice interfaces, 288
voluntary standards (*See* standard types)

Washington, D.C., 41
white background screens, 42, 43
WHO (*See* World Health Organization)
WI (*See* work item)
Windows, 281
work conditions, 277
work desk or work surface, 230
Work Environment Ordinance, 101
work item, 12, 13, 296
Work with Display Units Conference, 38, 40 (*See also* ergonomic conferences)
worker compensation
  association, 97
  claims, 55, 142
  insurance, 97
worker health and safety, 275
worker performance, 41
working groups, 12
Working Group, 296
workplace, 305
workplace standards, 16
workstation, 35, 305
workstations, 278, 280
World Health Organization, 20, 23, 40, 292
wrist
  disorders, 43
  injuries from unpadded rests, 240
  rest, 240, 305

X Windows-based environments, 128
Xerox, 272
X-ray act, 96

ZH 1/535, 36 (*See also* Germany)
ZH 1/618, 36, 37, 96 (*See also* Germany)